LIFE'S
ENERGY

LIFE'S ENERGY

STEFAN BRÖER

Library of Congress Control Number:		2022920175
ISBN:	Hardcover	978-1-6698-3261-4
	Softcover	978-1-6698-3260-7
	eBook	978-1-6698-3259-1

Print information available on the last page.

Rev. date: 11/07/2022

To order additional copies of this book, contact:
Xlibris
AU TFN: 1 800 844 927 (Toll Free inside Australia)
AU Local: (02) 8310 8187 (+61 2 8310 8187 from outside Australia)
www.Xlibris.com.au
Orders@Xlibris.com.au
844150

CONTENTS

1

The Energy of Life

While there is life there is hope. I beg to assert . . . that as long as a man's heart beats, as long as a man's flesh quivers, I do not allow that a being gifted with thought and will, can allow himself to despair.
—Jules Verne, *Journey to the Center of the Earth*

When we use the word 'energy', we use it in two different contexts. One is our psychological state; the other is energy as work and exercise. We often say, 'I feel energetic today' or we feel 'deflated and lack motivation'. This has relatively little to do with how our body generates energy to carry out work, or to energise our brain. Our body is always ready to generate more energy even if we feel deflated. Most of this book deals with not only the energy to carry out work and exercise but also the energy required to think, generate emotions, and let our organs work. For all of this, our body uses but one molecule. Despite its importance, it is not known in the wider population, but this book will put it centre stage.

This molecule is called ATP or by its proper name *a*denosine *trip*hosphate (or adenosine with three phosphates, simplified in Figure 1), which is to any biochemist the energy of life. Do not worry about the chemistry; this book will explain all concepts at a level suitable for any science-interested reader. I will not show chemical formulas throughout the book, but for ATP, I must make an exception because there is no other way to introduce it properly.

Adenosine is made up of a molecule called adenine linked to a sugar. We don't need its precise structure, but the adenine-sugar part serves as a key, so ATP can only enter places where the key fits. The business end of the molecule is the three phosphate groups, which provide the energy. When furnished with phosphates, it is called a nucleotide. In Figure 1, the phosphates are shown as a *P* in a circle. In case you asked, phosphate is a phosphorous atom bonded to four oxygen atoms. Phosphate is very stable and easily soluble in water.

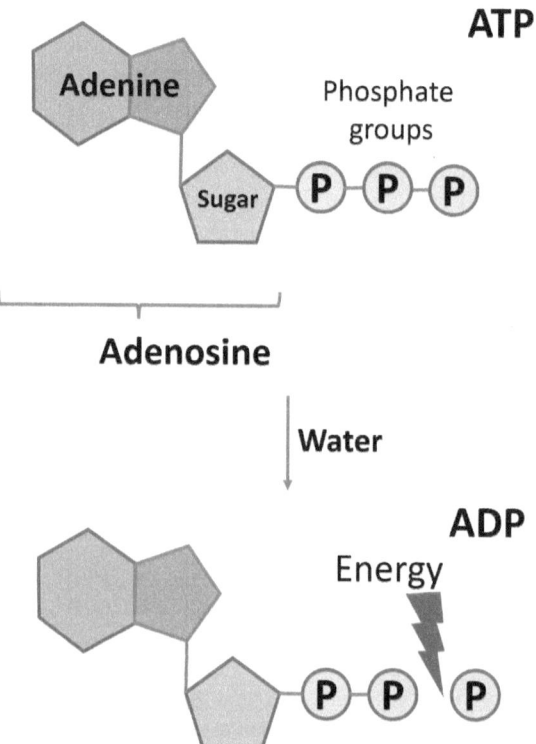

Figure 1. Simplified depiction of ATP. Phosphate is shown as a circle with a P. Adenosine is made up of sugar plus adenine. Splitting of a phosphate with water releases energy.

As an example, Coca-Cola contains about 0.2 g of phosphate per litre. The oxygen atoms can build a link between phosphate molecules, and this is how the three *P*s are connected in ATP. Water can be used to break this bond, and when this happens, energy is released. It is quite easy to break this bond; if the water is made acidic, it will happen spontaneously. This is important because it shows that the energy in this bond is readily released. Or in other words, the products are more stable than the ATP. When this happens, the shortened molecule is called ADP, which stands for adenosine diphosphate.

Most people know ATP as the Association of Tennis Players, which is a segue to my first example to illustrate the role of ATP. A tennis player uses about 23 kg of ATP per hour in a tennis match. This is quite astounding because the tennis player is not losing 23 kg body weight, but she is using all the energy provided by the release of 4.4 kg phosphate groups. In fact, she is only going to lose 60 g of body weight per hour, in the form of carbohydrates and fat. After the match, the balance will say that she lost more weight, but this is only water as sweat. This also explains why it is so disappointing to lose weight just with exercise. The difference between using 23 kg ATP and losing 60 g of stored nutrients suggests that ATP is constantly split into ADP and phosphate and quickly fused together again to reform ATP during the tennis game. Something is churning in her body, maintaining high energy levels for a match that can last three hours. How is this possible? We will explore the reactions that recycle ADP back to ATP in the next chapters.

Actually, our body is not that expensive to run. We only use as much as a 100 W light bulb when resting, but I think we are getting more value for money out of our body than from a 100 W light bulb. A professional cyclist, however, can produce another 300–400 W when cycling uphill or more than 1,000 W during brief bursts of activity.

An effective way to understand the role of ATP is by comparing it to money. I will use the economy as an analogy several times in this book because our body is a good bookkeeper of energy expenditure and intake. You can view ATP as cash. To stay close to the analogy, you can buy electricity or petrol with cash. This will provide you with energy that you can use for a lot of things, such as heating your house (keeping your body temperature), driving around (exercising your body), or running a computer (running your brain). One molecule of ATP is a small amount of cash, like a one-dollar coin (or one euro or whatever you prefer). In the economy, currencies go up and down in value, and so does ATP. Our body is desperate to keep it at one dollar, but during strenuous exercises, its value may go down to ninety or eighty cents in muscle. This initiates a lot of emergency responses to bring it back to one dollar. If we have a stash of ATP and ADP, we want ten molecules of ATP for each molecule of ADP. In that case, our ATP is worth one dollar. If this goes down to let us say eight molecules of ATP and three molecules of ADP, our body energy is running low already. Muscle is reasonably tolerant to devaluation; we just get fatigued and stop running, but not our brains or our hearts. A heart attack is so dangerous because our ATP is devalued very suddenly, and then the heart stops beating. If that happens in the brain, it is called a stroke, and our neurons stop working and die. We will come back to this later in much more detail, but it nicely illustrates the importance of ATP. As in real life, if we spend money, we need to refill our purse by having a job that provides us with a salary. In the case of ATP, that is called nutrition and is used to convert ADP back to ATP. In short, we need to eat to keep ATP up at one dollar. To be realistic, ATP is much smaller than a dollar. We may spend a couple of hundred dollars in a day, but 23 kg of ATP represents 2.7×10^{25} molecules. To give you an idea of the scale, one molecule of ATP compared to the total of 23 kg is like a shot of whiskey compared to the total volume of all oceans on earth. ATP is a very small coin indeed, and we are going through a lot of it because exercise requires a lot of energy.

What kind of energy is released when ATP splits into ADP and phosphate? It is vibrational energy.[1] Whatever the ATP is attached to will wobble and shake. Wobble and shake at the level of a molecule is the same as heating. When water boils, it is rolling because the water molecules bounce and wobble so much that they become a gas instead of a liquid. Splitting ATP is like a local heater bringing the molecule to 3900 °C.[1] This is the reason why ATP is such a small but powerful unit of energy. At a larger scale, everything would just be scorched, because the water would evaporate, and everything clumps together into an unrecognisable mess. Remarkably, humans turn over more energy per second than the equivalent weight of sun material.[2] When the heat of splitting ATP is applied locally, however, the wobbling will provide force to do something. It is also important that the wobbling is not transferred to water molecules in the surrounding because this would only heat up the water. To some extent, this is unavoidable and the reason we warm up when we exercise, but a good chunk of the energy is used to do work. To form ATP, the opposite must happen. We have to apply energy very locally to fuse ADP and phosphate together to make ATP.

As we will see throughout the book, the vibrational energy associated with the splitting of phosphate from ATP can be used for many things, such as contracting muscles, pumping ions, facilitating chemical reactions, and switching proteins on or off.

The relation between work, heat, and energy was established during the Industrial Revolution.[3] Particularly important was the observation that different forms of energy can be interconverted. The steam engine, for instance, converted heat into mechanical energy. Sadi Carnot (1796–1832) established that heat flowing from hot to cold could generate mechanical energy. James Joule (1818–1889) then demonstrated that electric currents could heat up water. These experiments showed that different forms of energy could be converted into each other, but the energy never disappears. These and related observations started a new branch of physics called thermodynamics, which was established through the minds of James Joule, Julius von Mayer (1814–1878), Lord Kelvin (1824–1907), Hermann von Helmholtz (1821–1894), and Rudolf Clausius (1822–1888). Julius von Mayer derived his ideas about conversion of energy from biological observation. As a doctor in the East Indies, he discovered that venous blood was much brighter in colour than in colder climates.[3] This means less oxygen was extracted from blood to generate cellular energy in the tropics. He recognised a link between work and heat production in humans and the ambient temperature. Higher ambient temperature reduced the amount of heat generation by the body and reduced the amount of oxygen used by the body. Thus, heat and work were equivalent forms of energy. Julius von Mayer never got the credit for his ideas during his lifetime because he formulated them only as concepts. Hermann von Helmholtz subsequently analysed the conversion quantitatively and developed the thermodynamical theory around it in 1847.

Although different types of energy can be interconverted, heat only flows from hot to cold but never in the opposite direction. This is the famous second law of thermodynamics. To illustrate the law, let us go back to our example of ATP wobbling when phosphate is cleaved off. It makes intuitive sense that the wobbling can easily be propagated to neighbouring water molecules, which than wobble a bit more as well. The wobbling (heat) is distributed to many more water molecules and therefore dilutes and spreads very quickly. As a result, the breaking of the ATP molecule heated the surrounding water by a small amount. The opposite is not intuitive. For that the random wobbling of water molecules would have to come together all at the same time and at the same point to bang a phosphate into ADP with sufficient force to make ATP. This is very 'unlikely' and explains the second law of thermodynamics, which states that heat energy dissipates very quickly (flowing from hot to cold). It is impossible that a select group of water molecules will receive wobbles from many other water molecules so that they start to warm up (wobble more) in one area, while the neighbouring areas cool down and wobble less. As the second law of thermodynamics implies, heat gradients in the universe will eventually all disappear at which point no convertible energy is left. Currently, the energy is highly concentrated in stars like our sun, which in turn provides the energy for life on our planet. The only way to locally bypass (not violate) the second law of thermodynamics is with information, which is the secret of life. We will come back to the secret of life in the last chapter. For now, we can say that there is more to ATP than just exercise.

As a biochemist, I would be so bold to say that ATP is a definition of life, at least on our planet. It could be a different currency on another planet, but on this earth, there is no life without ATP. ATP must be very ancient. You may not have thought about it, but we use a slightly modified ATP to make DNA, namely a deoxy-ATP[a] to make deoxyribonucleic acid. If you look at the famous double helix, its rungs and rails are made up of pairs of nucleotides (*A* and *T* form a pair and *G* and *C* form a pair). Guess what *A* stands for: adenosine, or adenine if you are just looking at the rungs. If you are now asking, 'Where are the phosphates?' you are on to something. When nucleotides, such as deoxy-ATP, chain together to form strands of DNA, two of the three phosphates are lost. One phosphate is retained and becomes part of the rails, connecting the sugar units. *G*, *C*, and *T* also start with three phosphates, losing two while getting connected. Thus, long strands of DNA can be made in our cells using the energy that is liberated when the two phosphates fall off. Ribonucleic acid (RNA) is made in the same way but uses unmodified ATP. RNA has become as famous as DNA because the first vaccines to prevent severe infection by COVID-19 are based on RNA, instructing our cells to make a part of the virus that forms the spike protein on its spherical surface. Life tends to reuse successful parts or building blocks, as biochemists like to call them. For instance, ADP (ATP after it lost one phosphate) is found in quite a few molecules in our body. As we will see later, they all play essential roles to provide us with energy all day.

---

[a] An ATP with an oxygen missing in the sugar part of the molecule.

ATP can even be used to generate light.[4] If you have seen fireflies in the summer, you have seen a reaction that converts the energy of ATP into light. This is called the luciferin and luciferase reaction after the two molecules that are involved in addition to ATP. When it was discovered, an aqueous extract of the glowing part of the insect was generated. It was literally firefly butts ground up. The extract would glow briefly and then stop. The process could be repeated by generating two extracts, a hot and a cold extract. If you added the two together, the mixture would glow again for a short while. This can be explained by the components of the reaction. The reaction is catalysed by the enzyme[b] luciferase, a protein. Proteins, like egg white or enzymes, are destroyed when heated. This is the reason clear and liquid raw egg white becomes white and solid upon heating. The cold-water extract contains the active enzyme, which uses up the ATP in the extract to make light briefly. When ATP is used up, the light dims. In the hot extract, the ATP is conserved, because all the enzymes that may use it were destroyed. Combining the two allows the reaction to resume.

I am outlining this type of experiment in some detail because early biochemists, whom we will meet in the next chapters, have used such experiments to understand how organisms generate energy. They called the mix of enzymes in the extracts ferment and heat-stable molecules that are used up in the reactions, such as ATP, coferments (we call them coenzymes nowadays).

But how does ATP generate light? Light is photons, which are generated when electrons in a molecule drop from a higher to a lower energy level. ATP provides the energy to allow the molecule luciferin to assume an 'excited' electron configuration. In subsequent reactions, the electron configuration is rearranged, the energy level drops, and light is emitted. Then the whole cycle can be repeated using another ATP.

---

[b]   An enzyme is a biological catalyst that lets a certain chemical reaction happen at body temperature which otherwise would be very, very slow.

This brief introduction gives you a preview what ATP can do and what this book is about. We will first go back in history and see how ATP was discovered. Then we will be following ATP around our body and see what it does. We will look also at its role in diseases. Along the way we will learn about the scientists involved in ATP research, how their research activity was affected by the rise of the 'Third Reich', and we will meet many of the Nobel laureates of the twentieth century.

2

A Missed Nobel Prize

Science, my lad, is made up of mistakes, but they are mistakes that are useful to make because they lead little by little to the truth.

—Jules Verne, *Journey to the Center of the Earth*

To appreciate the discovery of ATP, we must draw a wider circle to understand the development of life's energy principles. We often say that we want to 'burn' some calories. This is close to the truth. However, the burning of food in our cells is not a sudden combustion but rather a slow process that occurs in small steps.[5] As you know from your car – unless you bought an electric one – internal combustion engines require oxygen to work. This is the work of the carburettor. It generates a mist of petrol in air, the latter containing 20% oxygen. After ignition by a spark, the mixture combusts into carbon dioxide, water, and some minor less desirable fumes. Chemically speaking, petrol is quite closely related to fat. The molecules are a bit shorter, making it liquid, while fat is longer and chemically modified at one end, rendering the molecule a solid. When we burn foodstuff, we capture its energy during a stepwise breakdown of our nutrients until the final oxidation with oxygen to generate water and carbon dioxide. Generation of ATP is the main purpose of this process, as we will see later.

Joseph Priestley (1733–1804) was the first to isolate oxygen by heating mercuric oxide, which generates mercury and oxygen gas. He found that a candle burnt more brightly in pure oxygen and a mouse survived longer in this gas than in normal air. Priestley called it dephlogisticated air, holding on to an old theory that suggested that fire (phlogiston) was a substance released from combustible items.[c] At this point, Priestley visited Antoine Lavoisier in Paris (1743–1794, Figure 2) and told him about his experiments. Lavoisier immediately recognised the importance and repeated the experiments. He later claimed that he had discovered oxygen independently, generating a bitter dispute with Priestley.[3] In Figure 2, Antoine Lavoisier is depicted with his wife, Marie-Anne, who played a significant role in their research as a discussion partner, note-taker, and artist drawing the experimental equipment. To Lavoisier's credit, he realised that this gas was a new element that could combine with metals to form oxides and that animals and humans use it to burn nutrients to generate energy and heat. Although Priestley discovered oxygen, Lavoisier was the first to fully recognise the principle of oxidation and combustion.

He called it 'eminently breathable air' and later named it 'oxygen'. When a candle was placed in a jar, the oxygen disappeared, and a similar amount of carbon dioxide or 'fixed air' was generated.[6]

One of Lavoisier's greatest strengths was his meticulous bookkeeping and balancing of all reactions. He used precision scales, measured the amount of gas produced in a reaction, and measured the generation of heat from the rise of temperature of a water bath in which the reaction vessel was submerged.

---

c   Today we know that fire is the rapid oxidation of gases released from heated material such as wood or from materials with a low boiling point such as petrol.

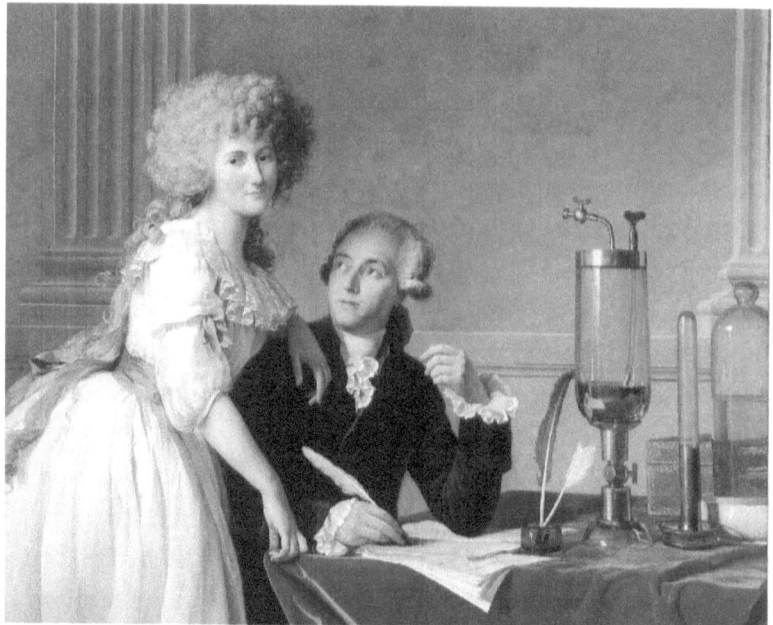

Figure 2. Portrait of Antoine Lavoisier and his wife, Marie-Anne. Painting by Jacques-Louis David, Public domain, via Wikimedia Commons. Note that Lavoisier looks at his wife, not the observer.

Lavoisier established the law of mass conversation, which states that in any reaction, the weight of all substrates of a reaction matches the weight of all products. This law explains why our tennis player did not lose weight during the game because 23 kg ATP combined with 0.8 kg water become 19.4 kg ADP and 4.4 kg phosphate, and the reverse happens when ATP is reformed. Using these experimental techniques, Lavoisier, together with the mathematician Pierre-Simon Laplace (1749–1827), showed in 1782–83 that a guinea pig consumed oxygen and produced carbon dioxide. Moreover, the amount of heat produced by the guinea pig was proportional to the amount of carbon dioxide produced. The amount of generated heat was similar when charcoal was burnt to produce an equivalent amount of carbon dioxide as in the guinea pig experiment. These observations lay the foundation of our understanding of animal and human metabolism.[6]

We can write the following:

Foodstuff (containing carbon) + Oxygen → Energy + Carbon dioxide

The conversion of nutrients inside the cell is referred to as its metabolism. For Lavoisier, heat was a substance, a massless fluid termed 'caloric' (from the Latin *calor* = heat). The theory was mistaken,[d] but we still use the term today to quantify the energy content of food in units of calories. Lavoisier, who was a tax collector in his day job, became a victim of the French Revolution and was beheaded because of his involvement in politics. The mathematician Joseph-Louis Lagrange (1736–1813) lamented that it took only a moment to sever Lavoisier's head but that it would take a hundred years to produce another one like it.[6]

While Lavoisier thought that oxidation (combustion) was taking place in the lungs, Georg Liebig (1827–1903), son of the more famous Justus whom we will meet in a moment, and Carlo Matteucci (1811–1868) then established around 1850 that oxidation occurred in tissues and that the blood was transporting oxygen to the tissues.[7] This was further elaborated by Eduard Pflüger (1829–1910) in 1872. The chemist Jean Baptiste Dumas (1800–1884) stated in 1841, 'An animal constitutes, in effect, a combustion apparatus, from which carbonic acid [carbon dioxide] is continuously disengaged, and in which, consequently carbon continuously burns.'[8]

Lavoisier established the basic principle of energy generation in organisms, but chemistry had only just started to become a mature scientific discipline and had not yet developed sufficient analytical skills to identify any of the components or reactions that occur inside a cell. This changed in the nineteenth century when Justus von Liebig (1803–1873) identified inosinic acid in muscle extracts in 1847.[9] Inosinic acid is a breakdown product of AMP, which is

---

[d]   As we saw earlier, heat is the extent of molecules wobbling and bouncing around.

ATP that has lost two phosphates (one step more than shown in Figure 1). AMP and inosinic acid spontaneously develop in muscle extracts from ATP.[e] As we will see, the instability of ATP delayed its discovery by almost one hundred years. Justus von Liebig (Figure 3) was an outstanding chemist and understood animal and plant nutrition, but biochemistry – the discipline that aimed to understand what happened to foodstuff inside a cell – did not develop until the twentieth century because of analytical limitations. The discovery of AMP was part of Justus von Liebig's work to catalogue the molecules that make up an organism.[10] He established the three main types of molecules that make up an organism, namely carbohydrates, fat, and proteins.

Around the same time, Theodor Schwann (1810–1882) isolated the first enzyme from gastric juices, which he called pepsin. Enzymes are biological catalysts that facilitate reactions, which at body temperature would otherwise proceed only extremely slowly. The combination of many enzymes allow the conversion of food molecules into carbon dioxide, water and energy. Enzymes are proteins, typically made up of hundreds of amino acids that form an intricately folded structure, but these insights would not be achieved until the twentieth century. Gastric juices could break down food like fermentation processes, and as a result, enzymes were called ferments. Justus von Liebig was heavily opposed to these views, believing that digestion was a purely chemical process. Schwann went further and proposed that all living organisms are made up of cells, which carry out the functions of each organ. The conversion of nutrients inside the cell, which is referred to as its metabolism, is exemplified by the fermentation of sugar into alcohol and carbon dioxide by yeast. Schwann wrote,[8] 'Cells . . . must have the faculty of producing chemical changes in its constituent particles. Besides which, all the parts of the cell itself may be chemically altered during the process of its vegetation. The

---

[e] Inosinic acid is still used as a flavour enhancer by the food industry. No surprise we like the taste of meat.

underlying cause of all these phenomena, which comprise under the term metabolic phenomena of cells, we will denominate the metabolic power.' Justus von Liebig ridiculed the cell theory and destroyed any career prospects for Schwann in Germany. As a result, he left Germany to become a university lecturer in Belgium.

Figure 3. Early pioneers of chemical physiology. Left: Portrait of Justus von Liebig by Franz Hanfstaengl. Right: Claude Bernard. Public domain, via Wikimedia Commons.

Justus von Liebig was also a scientific competitor of the French physiologist Claude Bernard (1813–1878) (Figure 3). Both had an intense interest to understand how foodstuff is digested and stored inside the body. Claude Bernard recognised that blood sugar is the main carrier of energy in animals and that animals can generate sugar even in the absence of carbohydrates in the nutrition to maintain blood sugar levels during fasting.[8] He also recognised that acidification accompanies muscle contraction through the generation of lactate. Moreover, Claude Bernard recognised the principle of homeostasis. Homeostasis is defined as the tendency to maintain a constant internal environment in organisms.[11] This is the reason we can use thermometers to find out whether we have an infection.

Homeostasis also lets us breeze faster when we exercise because we need to exhale more carbon dioxide produced by our muscles.

Moreover, we can go to the doctor and have a blood test done. This can reveal whether we have elevated blood glucose and may have an onset of type 2 diabetes. As a final example, we saw in the introduction that ATP levels have an extremely strict homeostatic control so that out of eleven molecules, ten are ATP and only one is ADP. We compared this to maintaining the value of a currency close to one dollar.

The key step in the development of biochemistry as a discipline and in the elucidation of cellular metabolism occurred with the discovery of Eduard Buchner (1860–1917) (Figure 4) that a yeast extract could carry out the fermentation of sugars to alcohol in an equivalent way as an intact cell.[12] Eduard Buchner generated the cell extract by grinding yeast with pulverised quartz using a mortar and pestle. He then used a press to squeeze out the cells' liquid content. This extract, however, quickly decomposed upon storage. Eduard's brother Hans suggested using sugar to preserve the extract, like the preservation of fruit juice in jam. Trying this, Eduard Buchner made the seminal observation that the extract started to bubble and produced carbon dioxide, like intact yeast would do.

Louis Pasteur (1822–1895) famously said, 'In the fields of observation chance only favours the prepared mind.' Buchner's conclusion that he had reactivated a complete metabolic process in a cell extract can be considered the birth of biochemistry as a discipline and a canny example of Pasteur's quote. The sugar was converted into alcohol and carbon dioxide. He published this observation in 1897 under the title 'Alkoholische Gärung ohne Hefezellen' (Alcoholic Fermentation without Yeast Cells).[13] From now on, pure chemicals could be added to cell extracts and their effect studied on the metabolic process under observation. The discipline of biochemistry or physiological chemistry, as it was called then, had started slightly earlier with the

Figure 4. Founders of biochemistry. Left: Eduard Buchner, official Nobel Prize photo from 1907. Public domain, via Wikimedia Commons. Right: Arthur Harden, 1927, Nobel Foundation. Public domain, via Wikimedia Commons.

appointment of Carl Gotthelf Lehmann as an adjunct professor of physiological chemistry at the University of Leipzig in 1843, who also wrote the first textbook of the discipline in that year.[8] However, before Buchner, physiological chemistry was largely concerned with cataloguing the chemicals of living organisms. Eduard Buchner received the Nobel Prize in 1907 for this discovery, not only because it opened a new discipline, but also because it put to rest the notion that living organisms had forces or principles distinct from those derived from chemical or physical laws. Proponents of this view, called vitalists, thought that cells contain a complex matrix called protoplasm, which carried out life's functions. Justus von Liebig was one of the most influential vitalists in his time, but Buchner's experiments were a complete vindication for Schwann's cell theory. The protoplasm remained a popular concept for some more years, until it was fully appreciated that processes like fermentation were carried out by a series of enzymes facilitating individual chemical reactions. In 1897, Buchner wrote, 'The agent responsible for the fermenting action of the press juice is rather to be regarded as a dissolved substance, doubtless a protein; this will be denoted zymase [later called enzyme].' In 1901, Franz Hofmeister (1850–1922) echoed this view saying, 'Every type of chemical reaction in the cell corresponds to a ferment [enzyme].'[8] As I mentioned above, pure chemicals could be added to this extract and their effect studied on the process of fermentation. Arthur Harden (1865–1940) (Figure 4) and William Young used this method in 1906 and added phosphate to yeast extract only to observe that the production of carbon dioxide increased.[14] Moreover, the amount of carbon dioxide was proportional to the amount of phosphate added. This was not yet a proof or discovery of ATP, but it was the first evidence that phosphate is an important inorganic molecule that plays an intricate role in the conversion of foodstuff – which is only composed of carbon, hydrogen, oxygen, and nitrogen – into energy and carbon dioxide. Consistently, Harden discovered sugar-phosphate compounds in the extracts but always maintained the view that these compounds were products of a side reaction in the breakdown of sugar into carbon dioxide and alcohol.[15] He was proved wrong on the latter claim but received the Nobel Prize in 1929 for his discoveries. His view can be understood because neither the starting compound glucose nor the final products of yeast fermentation, namely ethanol and carbon dioxide, contain any phosphate. Moreover, in a chemical laboratory glucose can be broken down without the involvement of phosphate.

Arthur Harden spent some time in the laboratory of Emil Fischer (1852–1919, Nobel Prize 1902) in Erlangen, before returning to Manchester. Emil Fischer was the eminent organic chemist of his time generating a lasting legacy through training of young researchers. In 1897, Arthur Harden became head of the Department of Biochemistry at the Lister Institute in London. He was a founding member of the British Biochemical Society.[16] By 1909, Phoebus A. Levene (1869–1940) and Walter A. Jacobs (1883–1967) had worked out that the inosinic acid, discovered by Liebig in muscle extracts, contained three different types of molecules and how they were linked together, namely a sugar, a phosphate, and a molecule similar to adenosine, in this case inosine.

At the same time, medically trained biochemists in Germany focused on muscle tissue to understand how energy is generated for muscle contraction. The first forty years of the twentieth century were a golden age of German biochemistry.[17] Otto Meyerhof (1884–1951), Gustav Embden (1874–1933), Otto Warburg (1883–1970), and Carl Neuberg (1877–1956) were key figures in the elucidation of cellular metabolism and energetics. The excellence of German biochemistry ended with the rise of the Nazis and had to be rebuilt after the Second World War. We will see many examples how biochemistry as a discipline moved from Germany to Britain and the United States because of a regime that placed racial considerations above science.

Otto Warburg, who was instrumental in the elucidation of the processes involved in cellular respiration, had a father of Jewish heritage and was proud to be German. In the First World War, he served in the Prussian horse guards and remained interested in equine sports throughout his life. Like Arthur Harden, he trained under Emil Fischer (1852–1919) and did his PhD under the guidance of Ludolf von Krehl (1861–1937). He lost his teaching position during the Third Reich but was allowed to continue his research in Germany because of his standing in the field.[18]

Otto Meyerhof (Figure 6) was born to Jewish parents in Hanover in 1884 and died in Philadelphia in 1951.[19] It was not immediately clear from his training that he would become one of the foremost biochemists of the twentieth century. Instead, he trained in psychology and philosophy. Meyerhof was a member of the neo-Frisian school of philosophy, which was founded by his friend Leonard Nelson (1882–1927). The school promoted critical philosophy, which became later a strong influence in mathematics. For many years, Meyerhof was the editor of a philosophy journal run by the neo-Frisian school. Meyerhof met his wife during his studies in Heidelberg. After graduation,

Meyerhof joined the laboratory of Ludolf von Krehl (1861–1937) in Heidelberg, who promoted the physiological basis of clinical medicine. In Krehl's laboratory, Meyerhof met Otto Warburg, who influenced his approach to science. Both scientists worked repeatedly together at the Marine Zoological Laboratory in Naples investigating sea-urchin egg metabolism. Otto Warburg turned Meyerhof's interest towards energy generation in cells. Meyerhof moved from Heidelberg to Kiel where he made his seminal discoveries on energy generation in muscle, which we will cover in a moment. Otto Meyerhof was still a research assistant in Kiel when he received his Nobel Prize in 1922. Surprisingly, the chair of physiological chemistry in Kiel was awarded to another person. However, Meyerhof was offered a position at the Kaiser-Wilhelm Institute in Berlin-Dahlem in 1924, where he met Otto Warburg again. In 1929, Meyerhof moved to the newly founded Kaiser-Wilhelm Institute for medical research in Heidelberg. The institute was founded at the initiative of Ludolf von Krehl. It was here that Meyerhof initiated the research that would result in the discovery of ATP. Meyerhof's laboratory was internationally renowned, and many of the scientists who worked with Meyerhof went on to become leaders in their fields, such as Fritz Lipmann (1899–1986, Nobel Prize in 1953, whom we will meet later), David Nachmansohn (1899–1983, whom we will meet as well), Hermann Blaschko (1900–1993, who discovered how neurotransmitters are synthesised), Severo Ochoa (1905–1993, Nobel Prize 1959 for the synthesis of RNA and DNA), George Wald (1906–1997, Nobel Prize 1967 for his research on pigments in the eye), and Andre Lwoff (1902–1994, Nobel Prize 1965 for the regulation of protein synthesis).[20] Being of Jewish descent, Meyerhof received a letter from Baden's Minister for Culture on 16 November 1935, which read,[21] 'In reply to your letter of November 15, I wish to inform you that the question of the maintenance of your honorary professorship has now been decided in the light of yesterday's implementing regulation in the negative.' Meyerhof's situation deteriorated quickly through the withdrawal of his teaching licence, forcing him to leave Germany in 1938 for Paris, where he was welcomed enthusiastically. In 1940, when France was invaded by Germany, he had to move on to the United States, where he found a permanent home at the University of Philadelphia. Since his PhD, he was accompanied by his wife, Hedwig, who survived him for several years.

Carl Neuberg, who is often considered the father of modern biochemistry, was also Jewish. He founded the *Biochemische Zeitschrift*, which published many of the seminal studies in this field in the early twentieth century. Like Meyerhof, he was forced out of his position in 1934 and had to rescind his position as an editor of the *Biochemische Zeitschrift*. Just before the outbreak of the Second World War, he moved first to Amsterdam and in 1940 to the United States, joining his daughters who had already settled there.

Gustav Embden was born in 1874 into a distinguished family.[22] He studied medicine in Freiburg and was introduced to biochemistry in Strasbourg under Franz Hofmeister (1850–1922). In 1904, Embden moved to Frankfurt and built up a state-of-the-art laboratory at the city hospital. In 1914, he became professor of 'vegetative physiology', the discipline that deals with the chemical functions of the animal body. In 1912, his group switched from working on liver metabolism to muscle metabolism. In contrast to William Harden, Embden developed the idea that phosphorylated sugar metabolites were integral to sugar metabolism. Embden also suggested that lactate formation was not directly connected to muscle contraction. Embden has been characterised as a romantic explorer, formulating bold ideas of which many had to be revised by experimental evidence and which resulted in controversies. At the same time, these ideas stimulated new research, and his final formulation of the breakdown of sugars in muscle cells has stood the test of times. Gustav Embden died in 1933. Jakub Parnas, whom we will meet in a moment, wrote in his obituary,[22] 'He died too soon and at an unhappy time in the history of science in Germany; he had not, however, himself to submit to the hardships and difficulties which were experienced by many of his colleagues.' The latter was not quite true, as Gustav Embden was humiliated by Nazi students as a Jew and was subsequently admitted to a nervous sanatorium where he died aged fifty-nine.

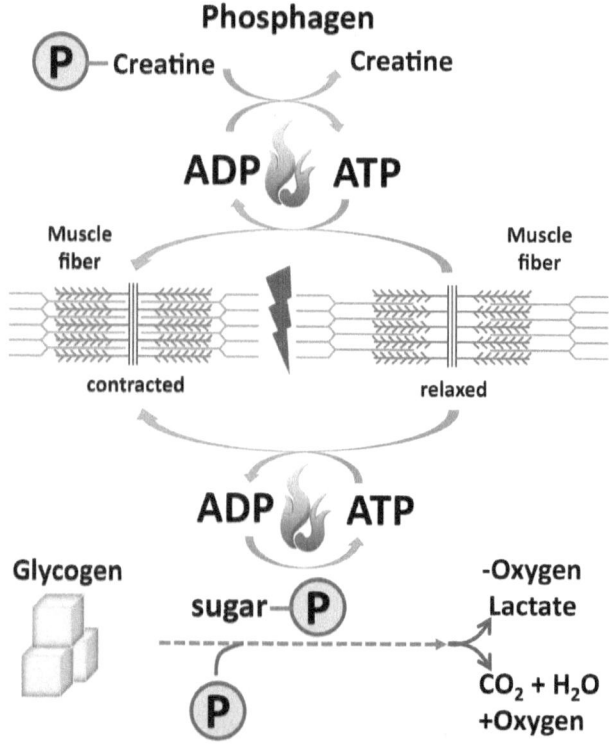

Figure 5. Energy generation in muscle. Sugar cubes under the label glycogen symbolise the stored glucose in muscle. Phosphate molecules are shown as a circle with a P. Fire symbols indicate the generation of heat. The lightning symbolises the electrical trigger for muscle contraction.

To understand some important developments in muscle biochemistry at the beginning of the twentieth century, I have drawn an overview of muscle energy generation in Figure 5. In the centre, we have muscle fibres relaxed on the right and contracted on the left. To contract, muscle filaments slide past each other to reduce the length of the fibre assembly. This movement occurs upon electrical stimulation by nerves, contacting the muscle tissue, symbolised by a lightning. The electrical stimulation itself does not cause the contraction; it just triggers it. Anyone who touched a live cattle fence has witnessed this powerful trigger. The sliding movement of the filaments requires ATP. ATP provides this energy by releasing one of its phosphates, generating ADP. Heat is generated in this process as well, symbolised as flames in Figure 5.

Many molecules of ATP do this simultaneously along the filaments, causing small arms to push the filaments along. This ATP action was discovered much later, a process that we will discuss in more detail in the next chapter. ATP can be regenerated from ADP by attaching a phosphate group with the input of energy. This can be accomplished by the breakdown of sugar, shown in the lower half of Figure 5, or by an energy storage system, called phosphagen (generator of phosphate), shown in the upper half. Phosphagen can be compared to a small battery that provides energy for a short while but needs to be recharged for continuous exercise. It is chemically known as creatine phosphate, which can donate its phosphate to ADP reforming ATP. Creatine can be purchased in nutrition shops that sell supplements for bodybuilding and exercise. The hope is to increase the capacity of the phosphagen system. Sugars for energy production can be derived from blood glucose, but muscle also has its own store of sugars called glycogen (sugar cubes in Figure 5). When sugars are broken down in the absence of oxygen or when oxygen is limited, lactate is produced. In the presence of oxygen, carbon dioxide and water are produced instead.

Otto Meyerhof and Archibald Hill (Figure 6) used excised frog muscle to understand the heat generation of muscle contraction and its relation to lactate production. Hill is often considered as the founding father of biophysics, a discipline that applies physical principles to living matter and uses physical instruments for its experiments. Isolated frog muscle can be stimulated electrically to contract, and this can be done in the presence and absence of oxygen.

In 1859, Emil du Bois-Reymond (1818–1896) had shown that a frog muscle stimulated to contract to exhaustion was strongly acidic, whereas an unstimulated muscle remained neutral.[8] Based on the earlier detection of lactic acid in muscle by Justus von Liebig, he assumed that the acidity was caused by accumulation of lactic acid.

Figure 6. Key figures involved in the elucidation of glucose metabolism. Left: Otto Fritz Meyerhof, Nobel Foundation. Public domain, via Wikimedia Commons. Right: Archibald Vivian Hill. Public domain, via Wikimedia Commons.

Subsequent experiments by Walter Morley Fletcher (1873–1933) and Frederick Gowland Hopkins (1861–1947, Nobel Prize 1929) in 1907 [23] established that during muscle contraction, glycogen (a polymer of glucose) disappeared while lactate was produced, particularly in the absence or limiting oxygen. When oxygen was reintroduced, the lactate disappeared. Meyerhof carefully analysed the quantities of lactate, glucose, and glycogen during muscle contraction in the presence and absence of oxygen. We will come back to this when we talk about ATP and exercise.

What is shown in Figure 5 would have contained a lot of black boxes for Meyerhof and Hill. Of the two stores of energy in muscle, namely glycogen and phosphagen, Meyerhof and Hill only knew glycogen as a polymer of glucose molecules that can be quickly broken down to provide individual sugar molecules for energy generation. Glycogen had already been discovered by Claude Bernard.[24] When a muscle is excised, the blood supply is severed, and sugar cannot be obtained from the circulation. The transport of oxygen is also limited but can be excluded completely, by enclosing the muscle in nitrogen gas. Yet the muscle tissue remains excitable for some time to carry out experiments. Meyerhof and Hill analysed the breakdown of glycogen during muscle stimulation and balanced it with lactate and energy production. They showed that the heat production of contraction was proportional to the lactate production, but the link between sugar metabolism and contraction was unknown.

Gustav Embden (Figure 7) at the same time had worked out that phosphate combined with sugar molecules during the process.[25] He could isolate two different sugar-phosphate compounds depending on whether metabolism was poisoned by sodium fluoride or not. As we will see on several occasions, the use of poisons is another important method to understand biochemical processes. One of the sugar phosphates was the same as identified by Harden in yeast; the other was different. Both were considered intermediates of lactic acid formation and therefore called lactacidogen (generator of lactic acid). At the time, Hill and Meyerhof did not know that phosphocreatine could provide energy for muscle contraction as well. More importantly, they also did not know that ATP was required to contract muscle fibres and that the resulting ADP could be recycled using phosphocreatine or sugar metabolism. Both showed that heat was generated when muscle twitched and that different processes were generating the heat in the absence and presence of oxygen.

Although they did not quite solve the problem of energy generation in muscle, their meticulous work to understand the energetic requirement of muscle contraction and its quantitative relation to the production of lactic acid earned both the Nobel Prize in 1923. In retrospect, the Nobel Prize seems almost premature but deserved as Meyerhof's laboratory was instrumental to work out the sequence of reactions in the breakdown of sugars in muscle cells in the 1930s, and Hill went on to make substantial contributions to biophysics such as understanding the binding of oxygen to haemoglobin.

Figure 7. Key figures involved in the elucidation of glucose metabolism. Left: Jakub Karol Parnas (Public domain, via Wikimedia Commons) and Gustav Embden (Public domain, via Wikimedia Commons).

The complete metabolic pathway to break down sugar in cells is now known as glycolysis or the Embden-Meyerhof-Parnas pathway. We have already met Embden and Meyerhof. The third person, Jakub Karol Parnas (1884–1949) (Figure 7), was a Jewish-Polish biochemist working at the university in Lviv (also known as Lwow or Lemberg). He also studied the breakdown of glycogen and later identified the first reaction in the glycolysis pathway that converts ADP into ATP.[26] For a while, it was known as the Parnas reaction but later received its proper biochemical name. Jakub Parnas became a communist activist but was falsely accused of espionage and died in the infamous Lubyanka prison in Moscow in 1949.

Figure 8. Discoverers of ATP Karl Lohmann (left, Berlin-Brandenburgische Akademie der Wissenschaften) and Cyrus Hartwell Fiske (right, academictree.org).

Phosphagen, the battery that maintains ATP for a short while, was discovered in 1927 by Philip (1903–1954) and Grace (1901–1970) Eggleton[28] and independently by Cyrus Hartwell Fiske (1890–1978) and Yellapragada Subbarao (1895–1948).[29] Fiske (Figure 8) and Subbarao (Figure 9) also chemically identified the phosphate generating system of muscle as the molecule phosphocreatine. In contrast to Meyerhof, they did not add acid to their muscle extracts, which preserved high-energy phosphate-containing compounds. In the introduction, I briefly mentioned that ATP spontaneously loses phosphates in acidic solution, and so does phosphagen. Thus, minor changes to procedures can result in new discoveries. The lability of the phosphate group illustrates that phosphagen has a high energy content, which can be used for muscle contraction. To acknowledge the discovery of phosphagen by the Eggletons, Archibald Hill authored an article in 1932 entitled 'The Revolution in Muscle Physiology'.[30] In this article, he also admitted that his own efforts had only partially solved the problem of energy generation in muscle: 'I am ready, as you will see, to bear my share of the blame for an imperfect theory.' Gustav Embden had already observed in 1926 that lactate accumulation after muscle stimulation was delayed.[31] This could now be explained by phosphagen being used first to energise muscle followed by the metabolism of sugar to lactate, or to carbon dioxide in the presence of oxygen. Moreover, the Danish physiologist Einar Lundsgaard (1889–1968) had shown that muscle could contract even when glycolysis was stopped by poisoning. He injected the poison into the leg muscles of rats. The rats could still run around for five to ten minutes before collapsing into a rigor mortis–like state. Creatine had been known for some time to be abundant in muscle, but its function and its phosphate-containing form had remained elusive, due to its lability. Lundsgaard's experiment suggested that in the absence of lactate formation, phosphagen provided the energy for muscle contraction. The labile nature of the second and third phosphate in ATP stymied efforts to identify the complete ATP molecule until 1929. In that year, Karl Lohmann (1898–1978, Figure 8), working in Otto Meyerhof's laboratory, found a way to protect labile phosphate-containing compounds in muscle extracts by precipitating organic compounds with specific salts. Using this technique, intact ATP was finally discovered[32] by Lohmann and independently by Cyrus Hartwell Fiske and Yellapragada Subbarao.[33]

Yellapragada Subbarao (1895–1948, Figure 9) or Yella, as most colleagues called him, arrived in Boston in 1923 after his medical training in India.[34] Because he had no licence to practice medicine in the United States, he worked initially as a night porter at Brigham and Women's Hospital.

He made friends at the hospital and became a researcher in the biochemistry department, where he discovered phosphagen and ATP. Despite his momentous achievements, he was denied tenure and recognition because of his reclusive personality. Instead, he joined Lederle Laboratories,[f] where he tried to isolate the vitamin

---

[f]   Lederle was an independent pharmaceutical company producing antitoxins and vaccines before it was acquired by American Cyanamid and later by Pfizer.

folate to treat anaemia. While he failed to isolate enough of the vitamin, he succeeded in chemically synthesising the vitamin with help from Harriet Kiltie, a young chemist at Lederle. As a bonus, these efforts created several folate analogues that were developed into the chemotherapeutic agent methotrexate which was introduced into the clinic by Sidney Farber (1903–1973). Under Subbarao's leadership, the antibiotic tetracycline was later discovered at Lederle. There are very few scientists who have made such significant contributions to research and health in a lifetime.

Figure 9. Pioneers of biochemistry. Gerty and Carl Cori (left, Public domain, via Wikimedia Commons) and Yellapragada Subbarao (right, Public domain, via Wikimedia Commons).

Back to the discovery of ATP, which initially was seen as a technical advance and was not immediately followed by an understanding of its physiological role. When Lohmann presented his discovery at an international conference in 1929, it did not make any impact. In his 1932 article, Archibald Hill states,[30] 'The 'organic' phosphate [in muscle extracts] was not mainly a hexose ester [a sugar-phosphate compound], it was not the source of lactic acid [lactacidogen], but was largely adenyl-pyro-phosphoric acid [ATP]. I wonder whether we are still failing to see something which in ten years will seem obvious [namely that ATP provides the energy for muscle contraction and that phosphagen acts as a short-term battery to maintain ATP levels]?' (Explanations in brackets added by the author). The statement was based on experiments performed by Gerty (1896–1957) and Carl Cori (1896–1984) (Figure 9) who wanted to find out which compound in muscle released most of the phosphate and found that it was ATP. Gerty and Carl Cori would receive the Nobel Prize for their related work on glycogen in 1947. Hill's statement showed that two years after the discovery of ATP, its physiological role was not yet appreciated. Lohmann, however, came a bit closer to the 'obvious' role of ATP in phosphagen synthesis and glycolysis in a publication with Meyerhof in 1931.[35]

> The present experiments lay the foundation of the thesis that the synthesis of creatine phosphate [phosphagen] can take place . . . at the expense of ATP breakdown, whilst the resynthesis of ATP out of adenylic acid and phosphate [AMP and phosphate, ADP had not yet been appreciated] is made possible through the energy of lactic acid formation. One may also assume here a connection between the resynthesis of ATP and the hexose [sugar] phosphates as the source of phosphate. This would at once make understandable how ATP can act as a coenzyme of lactic acid formation. (Comments in brackets added by the author)

Two years later, a scheme for the chemical reactions of the breakdown of glucose in muscle (glycolysis) was proposed by Gustav Embden[36] and confirmed by Meyerhof.[37] Embden also recognised that the inosinic acid discovered by Liebig was a breakdown product of adenylic acid (AMP).

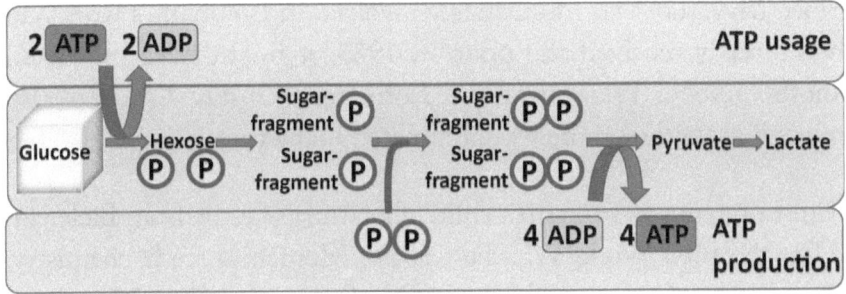

Figure 10. An overview of glycolysis highlighting the role of ATP. Phosphate is shown as P *in a circle.*

The scheme is quite complicated (Figure 10) because ATP initially transfers its phosphates onto sugar compounds forming the above-mentioned hexose phosphates or hexose esters, which after breakdown into smaller fragments donate the phosphate back to reform ATP. This explains the role of ATP as a coferment of the breakdown of sugar into lactic acid.

However, this by itself would not generate any extra ATP for other functions such as muscle contraction, because the same number of phosphates is added and given back. The net gain of ATP comes from the phosphate that Harden and Young found to accelerate glycolysis in yeast and is also used in muscle. The phosphate is attached to the sugar fragments and after chemical processing it is used to make ATP from ADP.

Lohmann proposed the correct chemical structure of ATP in 1935, but Katashi Makino had proposed the same structure a couple of months earlier [38]. The ultimate proof of the structure of ATP arrived in 1945 when it was synthesised by Basil Lythgoe (1913–2009) and Alexander R. Todd (1907–1997) in the laboratory and shown to be identical to ATP from tissues. Alexander Todd received the Nobel Prize for the synthesis of 'coferments' or to use the more modern word 'coenzymes' in 1957.

Since Meyerhof was immediately involved in Lohmann's work and had already received the prize in 1923, it might have prevented another Nobel Prize award to Lohmann. Gustav Embden also missed out, although he was nominated many times.

Karl Lohmann was the fifth child of a farmer's family from Bielefeld. After the First World War, he went to Munich to study chemistry. He moved to Göttingen for his PhD before joining Otto Meyerhof's laboratory. Lohmann moved to Humboldt University in Berlin in 1937 but never became a member of the Nazi Party. He headed the Institute for Physiology and Chemistry for fourteen years. In 1952, he moved to the Biochemistry Institute at the Medical & Biological Research Center of the German Academy of Sciences, where he continued to work even after his official retirement in 1964. In Berlin, his opportunities for conducting research were far more limited than in Heidelberg, and he devoted himself more to lecturing. For decades, Karl Lohmann, who would remain apolitical his entire life, represented the star biochemist of communist East Germany.[21]

In 1938 and 1939, Meyerhof's laboratory, and separately Negelein and Brömel,[39] unravelled the step during the breakdown of sugars that incorporates phosphate, thus explaining the acceleration of glycolysis in yeast extracts by phosphate as observed by Harden and Young in 1906. Moreover, Negelein and Brömel showed that in the next metabolic step ATP is produced, by the transfer of phosphate from an intermediate of glycolysis onto ADP forming ATP. Glycolysis has two of those steps (Figure 10). The first one was discovered by Jakub Parnas in 1934 and was hence called the Parnas reaction.[26] The second enzyme was isolated by Theodor Bücher (1914–1997) in 1942 who gave it its formal name. These two reactions established glycolysis as a pathway that generates ATP.

It took another ten years before it became clear that ATP was the direct source of energy for muscle contraction. Karl Lohmann showed in 1934 that creatine phosphate could recharge ADP to ATP, thus acting as an energy buffer in muscle.[40]

In 1939, Militsa Nikolaevna Ljubimova and Alexandrovich Engelhardt (1894–1984) showed that the major contractile protein of muscle, myosin, was intimately associated with an activity that splits ATP. The direct use of ATP for muscle contraction was then discovered independently by Albert Szent-Györgyi (1893–1986) and Joseph Needham (1900–1995) in 1942 (Figure 11).

Figure 11. Discoverers of the role of ATP in muscle contraction: Albert Szent-Györgyi (left, Public domain, via Wikimedia Commons) and Joseph Needham (right, Public domain, via Wikimedia Commons).

ADP, by contrast, was found to be incapable of eliciting muscle contraction. Albert Szent-Györgyi received the Nobel Prize in 1937 for the discovery of vitamin C and his work on metabolism. Around the same time, Fritz Lipmann (Figure 12) and Alexandrovich Engelhardt started to appreciate the turnover of ATP for metabolic processes. Lipmann summarised the field in a classic article,[41] in which he envisioned the breakdown of foodstuff through a "metabolic wheel" which generates a 'current' of phosphate-containing metabolic intermediates (~P). The squiggle ~P symbolises that the phosphate bond of these intermediates has high energy and can be used to make ATP from ADP. Creatine phosphate acts as a buffer. ATP is then used for metabolic processes of which he knew very little. Lipmann stated in 1941,[41] 'Not very definite answers can be given to the question as to how the high phosphate group potential [the high energy of the phosphate bond] operates as the promoter of various processes although a more or less loosely defined interconnection with phosphate turn-over is recognizable' (comments in brackets added by the author). Regarding muscle contraction, Lipmann said one year before Szent-Györgyi's discovery, 'The breakdown of creatine-phosphate, or more recently, ad-ph~ph~ph [ATP] are considered as the chemical reactions nearest to muscular contraction proper' (comment in brackets added by the author). Thus, Fritz Lipmann finally answered in 1941 Archibald Hill's question from 1932: 'I wonder whether we are still failing to see something which in ten years will seem obvious,' namely that energy from the breakdown of nutrients is used to attach a phosphate to ADP to form ATP, which is then used for muscle contraction and other energy-demanding tasks, converting it back to ADP and phosphate.

Fritz Lipmann received the Nobel Prize in 1953, not for his work on ATP but for his work on how organic acids are activated in metabolism, a process that requires ATP.

Fritz Lipmann (Figure 12) was born in 1899 in Königsberg but studied medicine in Munich and briefly served in the medical service in the First World War.[42] After the war, he returned to Königsberg, where he witnessed the Spanish flu. Lipmann became interested in biochemistry through a course in modern biochemistry in Berlin and through a brief stipend which led him to the University of Amsterdam. He decided to learn more chemistry in Königsberg in the laboratory of Hans Meerwein (1879–1965). After his medical exam, he joined the laboratory of Otto Meyerhof in Berlin. Through his brother and a friend, Fritz Lipmann had intensive contacts to the art scene, where he met his future wife, Freda.

Fritz Lipmann moved with Meyerhof to Heidelberg and later back to Berlin, where he received a fellowship of the Rockefeller Foundation, which he used to visit the United States from 1931 to 1932 with his now-wife Freda, before going back to Copenhagen where his mentor Albert Fischer had moved from Berlin. Alarmed by the growing influence of Nazi Germany in Europe, Fritz Lipmann moved from Copenhagen to the USA in 1939. He eventually settled down at Massachusetts General Hospital.

Figure 12. Early researchers of ATP generation and turnover. Herman Kalckar (left, academictree.org) and Fritz Lipmann (right, Public domain via Wikimedia commons).

Another biochemist who worked on the mechanisms that regenerate ATP in living cells was Herman Kalckar (1908–1991, Figure 12). He discovered that active respiration in the presence of oxygen can generate ATP by a different process than that described above in muscle in the absence of oxygen.[2] Comparable observations were made by Vladimir Alexandrovich Belitser (1906–1988) and E.

T. Tsybakova in 1939.[43] We will encounter this process in detail in the next chapter. Similar to Fritz Lipmann, Herman Kalckar summarised the principles of cellular energy production in 1941 in an equally influential article.[44]

In 1940, Severo Ochoa (1905–1993) demonstrated that ATP was required for the initial steps of glycolysis[45] (Figure 10) producing the sugar-phosphate compounds that had been discovered by Arthur Harden and Gustav Embden. In 1943, David Nachmansohn (1899–1983) identified the first biochemical reaction that required ATP as an energy source outside the glycolytic pathway. In this pathway, the neurotransmitter acetylcholine, which triggers muscle contraction, is generated by a series of reactions from acetate, one of which requires ATP as an energy input. We will meet acetylcholine again in the next chapter.

David Nachmansohn vividly describes the vibrant scientific community in Berlin in the late 1920s.[17]

> The KWI (Kaiser-Wilhelm Institutes) were located in Dahlem in a fashionable suburb of Berlin, built on lovely grounds and surrounded by lawns and gardens. They were purely research institutes founded in 1910. Large endowment funds had been provided by industrialists and bankers who realized that Germany's wealth was based on the rapid development of basic science. Chemical, pharmaceutical, electronic, optical, and many other science-based industries had transformed Germany in a half-century from one of the poorest to one of the wealthiest countries. Germany's economic wealth and power would greatly benefit from the strong support of scientific research. The members of the Institute had no teaching obligation. Therefore, in contrast to the Universities, no special fields had

to be represented. The selection of a director of an Institute or a subdivision was based exclusively on his scientific stature, competence, and the excellence of his achievements. This accounts for the extraordinary collection of brilliant scientists in a small area in about six Institutes. In the late 1920s, the KWI in Dahlem were one of the foremost and most outstanding scientific centres of Europe. . . .

Three such outstanding laboratories [Meyerhof, Warburg, and Neuberg] would have offered ample inspiration to any young biochemist. But one of the remarkable features of the KWI was the deliberate efforts, under the leadership of Fritz Haber, to break down the barriers between physics, chemistry, and biology. . . .

The famous 'Haber Colloquia,' which became a great attraction and played an important role in the activities of all the KWI. . . . In these seminars, Haber's extraordinarily brilliant mind was at its best. He stimulated vigorous and exciting discussions. His unusual ability to recognize and grasp the essential aspects of the many different topics presented, even if they were not in his field, to discover weaknesses and to raise pertinent questions, led to lively exchanges and sometimes to vigorous fights among the many outstanding people present. All these factors helped create a unique and exciting scientific atmosphere and made these seminars an unforgettable experience.[17]

Figure 13. Key figures working out the role of cofactors in biochemical reactions. Hans von Euler-Chelpin (left, Public domain, via Wikimedia Commons) and David Nachmansohn (right, National Library of Medicine).

David Nachmansohn (Figure 13) was born in Russia, but his parents moved to Berlin before he started school. He first trained at the Charitè, the major academic hospital of Berlin, and then joined Otto Meyerhof's laboratory. Being of Jewish origin, Nachmansohn moved to the Sorbonne in 1933. During this time, he attended several meetings of the British Physiological Society and became interested in the biochemistry of acetylcholine, the chemical that triggers muscle contraction. In 1939, he moved to Yale University and later to Columbia University.

Over the years, many ATP-dependent reactions were discovered that will form the basis of this book. One of these reactions, the light-emitting luciferase/luciferin reaction, I mentioned in the introduction. ATP was recognised as the energiser of this reaction in 1947 by McElroy.[4]

I want to leave the path to the discovery of ATP at this point because I will outline the role of ATP in many metabolic processes in subsequent chapters. From the Harden Young experiment in 1906, which was the first experiment to show that phosphate was required for the breakdown of sugars, it took thirty-five years to appreciate the role of ATP in metabolism as a coferment and as the power provider that drives many if not all cellular functions. Yeast and muscle played a key role, and the frog had to serve as the experimental hero. What I hope the reader will appreciate is that all researchers mentioned here – and quite a few that I did not mention – contributed pieces to assemble a puzzle of life's energy budget turnover. For good reasons, many of the protagonists received Nobel Prizes (Table 1), but many key discoveries also missed out, particularly the discovery of ATP itself. The discovery of ATP illustrates nicely that scientific discovery moves forward through the effort of a whole community. Even the best researchers misinterpreted their results and had to modify their ideas in the view of new experiments. What looks logical and convincing in hindsight is much more difficult to see at the time. A particularly vexing problem was the understanding of 'cozymase' or 'coferment', an elusive mix of compounds including ATP that are required for glycolysis to run.[43] Hans von Euler-Chelpin (1873–1964, Figure 13) isolated a compound that was required for the breakdown of sugar into alcohol in yeast and into lactate in muscle. This small molecule that was required for metabolism to occur was called cozymase by von Euler-Chelpin or coenzyme by Harden. The compound contained adenine, a sugar and phosphate. Although these components also occur in ATP, he had identified another 'coenzyme' which was later called NAD (nicotinamide adenine dinucleotide, in case the reader asks). A closely related coenzyme or coferment (NADP) was isolated by Otto Warburg in a heroic effort. To isolate and identify this coferment, Otto Warburg started with 100 litres washed horse erythrocytes, burst them by the addition of 200 litres of water, and precipitated protein by the addition of 500 litres acetone. This yielded 4.8 g coferment to elucidate its chemical composition.[46] Hans von Euler-Chelpin was awarded the Nobel Prize in 1929. Otto Warburg received the Nobel Prize in 1931 for his breakthroughs in the understanding of cellular respiration, which we will visit in the next chapter. Whenever a yeast or muscle extract was prepared, ATP quickly decomposed into AMP and NAD was used up. It took a long time to recognise that 'cozymase' or 'coferment' was not a single compound but a mixture of ATP, NAD, phosphate, and magnesium, all of which are essential for the reactions of glycolysis. Initially, cozymase was discovered similarly to ATP in the luciferase/luciferin reaction by generating two different extracts. One that contained all enzymes but had used up its ATP and other cofactors, and another in which all enzymes were inactivated by heat treatment, while the cofactors remain unaltered. Combining the two allowed glycolysis – or other reactions – to proceed for a while.

It is insightful to compare the elucidation of glycolysis and the role of ATP to a jigsaw puzzle, but without an image of the complete assembly. Initially, obvious structures can be identified resulting in a small facet. As the assembly goes on, bigger pieces join up, and sometimes it is simply hard work to place the final pieces with few structures to help. Lohmann could not have discovered ATP without all the prior steps and insights, and a full appreciation of all of ATP's functions took many more years. In the following table, I have summarised the main steps in the discovery of ATP in chronological order and the corresponding Nobel Prizes.

Table 1: Timeline of the discovery of ATP and related metabolic processes.

Year	Name	Discovery
1782	Antoine Lavoisier	Combustion as a source of energy in animals
1874	Justus von Liebig	Identification of inosinic acid in muscle extracts
1878	Claude Bernard	Homeostasis principle, glycogen as a store of glucose
1897	Eduard Buchner	Alcoholic fermentation in yeast extracts Nobel Prize 1907
1906	Arthur Harden William Young	Phosphate is required for the breakdown of sugars in yeast, Nobel Prize 1929 (Harden)
1908	Arthur Harden	Identification of sugar-phosphate esters in glycolysis
1923	Archibald Hill Otto Meyerhof	Nobel Prize for unravelling the energy generation during glycolysis
1927	Gustav Embden	Identification of phosphorylated intermediates of glycolysis and of AMP in muscle
1927	Philip Eggleton Grace Eggleton Cyrus H. Fiske Yellapragada Subbarow	Discovery of phosphagen or phosphocreatine
1929	Karl Lohmann Cyrus H. Fiske Yellapragada Subbarow	Discovery of ATP called adenylic acid pyrophosphate
1933	Gustav Embden	The first reaction scheme of glycolysis proposed
1934	Jakub Parnas	Identification of the first ATP generating reaction in the glycolysis pathway

1935	Katashi Makino Karl Lohmann	Structure of ATP proposed
1937	Herman Kalckar	Identification of a separate oxygen-driven ATP generating process persisting when glycolysis is blocked
1939	Erwin Negelein Heinz Brömel	Identification of a second ATP generating reaction in the glycolysis pathway
1939	Vladimir Belitser	Independent identification of the ATP generating process requiring oxygen
1945	B. Lythgoe Alexander Todd	Structure of ATP confirmed by chemical synthesis, Nobel Prize in 1957 (Todd)

3

Around the Arc de Triomphe

*Before all masters, necessity is the one most
listened to, and who teaches the best.*
—Jules Verne, *The Mysterious Island*

Anyone who exercises at a competitive level knows about lactate[g]
accumulation in muscle, particularly during high-intensity exercise.
As we saw in the previous chapter, lactate production was a key
indicator of muscle energy generation before ATP was discovered. In
the experiments described there, the muscle was excised and hence
lacked blood circulation, which exacerbates lactate production, due
to the lack of oxygen. The amount of lactate accumulated in muscle
during high-intensity exercise is a competition between blood flow
and ATP demand by muscle. The blood flow is not fast enough to
bring enough oxygen to the muscle tissue to avoid the production of
lactate. As we will see later, the most efficient way of ATP production
requires complete oxidation of nutrients to carbon dioxide and water
(combustion) as observed by Lavoisier in live animals. When oxygen
is not available in sufficient quantities, metabolism has a trick up its
sleeve, namely fermentation. During glycolysis, one intermediate is
oxidised not by oxygen, but by the removal of electrons, which for

---

[g] I will use lactate and lactic acid interchangeably. Technically one is the acid,
the other one the remainder when the acid-forming proton is removed.

a chemist is the same thing. The removal is mediated by Hans von Euler's or Otto Warburg's coferment (NAD). However, it can only store tiny amounts of electrons; and as a result, the electrons have to move on. As an easy solution, we put them back into the final product of glycolysis, generating lactate. This is a nice trick that works like a credit card or Afterpay. You can spend money faster than you earn it, but you cannot do it forever. As Gerty and Carl Cori (Figure 9) showed, balancing the credit card is done by the liver, which converts lactate back into glucose during the rest after exercise. If you are lucky in biochemistry, a particular part of metabolism is named after you. In this case, the shuttle of lactate between muscle and liver is known as the Cori cycle, first proposed in 1929.[47]

Due to the lactate credit card system, a short sprint is more a biomechanical problem than an energetic problem. Sprinters just run as hard as they can, and the energy reserves in muscle (ATP and phosphagen) are sufficient for 10 sec. Watch a sprinter after a 100 m run at the Olympics; they are not out of breath and panting. We will extend on energy reserves in a moment, but this happens all unconsciously to the athlete who just tries as hard as he/she can. A marathon run is something completely different. Here are some experiences from a marathon runner posted on the internet.[48]

For the first 20 km, he took an average of 4.5 min to run a kilometre. At the 10 km point, his legs still felt fresh, and his heart rate was in the mid-160s. He used the water stations and consumed glucose energy gels because glucose is more readily used by muscle than fat. After 30 km, the first un-ignorable signs of fatigue became apparent. The hamstring in his left leg tightened up a little longer than it should have. Changing the running style did not help as other muscles started to tighten up in the same uncomfortable way. He felt that his fuels were getting low. Despite consuming energy gels, he 'hit the wall' after 32 km, and the average time to run a kilometre increased to 5 min. When runners 'hit the wall', they refer to a physiological phenomenon that dictates a new mental and physical state to which the athlete has to switch. This change is not something that the runner can choose to ignore or push aside like they would any other type of pain during a long run or muscle tightness. At that moment, an athlete's body runs out of glycogen stored in muscle. It also affects the brain, because blood glucose levels drop slightly, and the brain can only use glucose but not fat. It takes time to release fatty acids from fat and turn protein into glucose, the fuels our brain and muscles can use. But muscle cannot generate the same amount of energy per second with external fuels as compared to endogenous glycogen. When an athlete pushes their body to the extremes of endurance, there is a point where all the 'easily utilised' energy – glucose from glycogen – is gone, and the body must now use glucose from the gels and begin to supplement fat to continue the activity. This is not an easy feat during a long-distance, strenuous activity. The runners' legs felt as if 10 kg weights were strapped to each ankle while trying to run at the same pace as before. The athlete tried to increase his pace back to 4.5 min/km, but his legs would simply not do it. Regardless of how hard he ran, his legs would not move any faster.

After 34 km, the pain was constant, and his legs and brain screamed to stop, despite the intake of energy gels. At this point, the marathon became a mental battle, but at a slightly slower pace, he could continue running. The lack of energy now caused the runner to swerve, and extra mental energy was required to stay on course. At 37 km, a small overpass made the runner's legs feel as if someone was stabbing a dagger into his quadriceps each time he took a step. The last 5 km were slightly downhill, bringing reprieve.

At this point in a marathon, the actual act of running became secondary to the mental capacity it takes to stay in the race. Right at the finish, the runner's legs begged him to stop, his lungs struggled to keep up, and his heart rate spiked from 175 to nearly 190 beats per minute. Collapsing at the finish line, he was hypoglycemic,[h] unable to walk in more than a shuffle, but elated about his achievement.

From this report, we can see that there is an energy crisis after three-fourths of the marathon, which we need to explore in more detail. However, with mental power, an athlete can move on and run much farther distances at a slower pace. Dean Karnazes, for example, ran 563 km in a little shy of 81 hours without sleeping in 2015 but, of course, had to consume nutrients while running.

The generation of energy for a sprint, where you do not even have to breeze, must be quite different from energy generation during a marathon. As we saw, the limits of a sprint are more biomechanical, and the limits of a marathon are more mental than energetic. The only energetic difference is that you cannot run a marathon at the same speed as a sprint.

If you think a human marathon is an exhausting experience, just imagine a fly with its high-frequency wing beat. It may seem like a cruel experiment to some readers, but you can glue the back of a fly to a metal pin with wax and lift it off the ground. At that moment, it will start to move its wings because the feet are not touching the ground. This 'flight' will go on until exhaustion. Interestingly, the fly has similar problems as the marathon runner. It has fat storage and glycogen storage. In 1949, Wigglesworth[49] examined the exercise capacity of one-week-old fruit flies, which the reader can find during summer in a fruit bowl or wine glass. These could fly for five to six hours before exhaustion. Upon exhaustion, the glycogen reserves had disappeared, but the fat reserves were still intact. When a 10% glucose solution was offered to the exhausted flies, they consumed it quickly and were ready to fly again. Remarkably, consumption of glucose solution for half a minute was sufficient to let the fly go for another 30 minutes. This tells us that carbohydrates are a readily

---

[h] Low in blood sugar.

available energy source, while fat needs to be mobilised first before it can be used.

Before we can appreciate the differences between a sprint and a long-distance run, we must understand how ATP is made in our body when oxygen is available. This is different from the experiments that Meyerhof performed at the beginning of the twentieth century. His muscle preparation was separated from blood vessels, and as a result, little oxygen reached the fibres. In that case, lactate was produced. When oxygen is brought to the muscle via red blood cells, the final product of glycolysis, a compound called pyruvate, is moved into the mitochondria. Mitochondria are the power plants of our cells, and most cells contain several hundreds of them. Mitochondria look a bit like bacteria and were indeed bacteria in the early stages of evolution, which were engulfed by larger cells that could only generate energy through fermentation. This mutually beneficial relationship, named symbiosis, became more intimate to the extent that a modern cell cannot live without them.[50]

The first person to call mitochondria the 'power plants of the cell' was Albert Claude (1899–1983) who used electron microscopy and developed separation techniques to study the components of cells.[51] He received the Nobel Prize in 1974 for his studies of cellular substructures.

Inside the mitochondria, carbon dioxide is generated from the nutrients we eat. The slow combustion that Lavoisier observed occurs in these organelles.[i] Without going into too much detail, the pyruvate molecule is taken apart, and each of its carbon atoms becomes carbon dioxide. In nutrients, chains of carbons are bonded to hydrogens or a mix of hydrogens and oxygen, while in carbon dioxide – as the name suggests – a single carbon atom is linked to two oxygen atoms. This

---

[i]   An organelle of a cell is similar to an organ of an organism. It performs a particular function that is important to the cell. Because it is on a much smaller scale, they are called organelles.

is the maximum number of oxygens you can attach to a carbon atom when fossil fuels or nutrients are burnt. The carbon chains must be broken down to allow two oxygens to bind to one carbon. Burning is the same thing as oxidation, chemically speaking, where as many oxygens as possible are attached to carbons, always generating carbon dioxide. Attaching oxygen to carbon is like pulling electrons away from carbon towards oxygen. This configuration is very stable, and therefore, energy is released in the process. If the concept of oxidation or removal of electrons appears strange to you, consider iron. Iron as a pure metal looks grey, but when it is oxidised and rusts, it becomes brown and oxygen atoms are attached to it. This change of colour is caused by the removal of electrons from iron towards oxygen. It is energetically favourable, so it happens spontaneously unless you cover the iron with paint. Iron is also involved in the transfer of electrons and the transport of oxygen in blood and muscle. We will come back to that in a moment.

We understand now that mitochondria are the places where carbon dioxide is produced, where we burn calories. I mentioned that in nutrients, the carbon atoms are surrounded by hydrogen atoms; where do they go? They are also oxidised, becoming water (H_2O). In water, two hydrogen atoms are bonded to one oxygen. That is also the maximum number of oxygen you can combine with hydrogen (one-half of an oxygen per hydrogen). We cannot attach two oxygens to one hydrogen, because there are not enough electrons. As a result, we must modify our nutrient-burning equation to add water:

Foodstuff (containing carbon, oxygen [few] and hydrogens) + Oxygen → Energy + Carbon dioxide + Water

Lavoisier had already recognised that exhaled air contained more water the ambient air. Water is also generated when fossil fuels are burned, which we can witness as water coming out of the car exhaust pipe. This is obvious in winter when the water condensates into white puffs. The reader can also breathe against a mirror to see water vapour coming out of our lungs. Tiny as mitochondria are, water and carbon dioxide are generated in two separate places, and this is the whole trick how we avoid rapid combustion of our nutrients.

Instead of just generating heat, we want to move our muscles. To do so, we store the energy in ATP instead, but how? For that, we need to look a bit closer at what happens to our nutrients inside mitochondria. To break down the carbon chain of the molecules (nutrient fragments) that move into mitochondria, we need to remove the hydrogens, and this allows us to break the molecules apart. Hydrogens are protons with an accompanying electron; and there are specific vitamins or derivatives of vitamins that can do just that, remove protons and electrons, typically two at a time and store them. We have come across these molecules already as Hans von Euler's and Otto Warburg's 'coferment'. If the reader has ever wondered why vitamins are good for us, here is the answer. They are 'coferments' participating in metabolic reactions.[j]

To see how the breakdown of molecules works, I will use an analogy. Imagine a large roundabout, perhaps the one around the Arc de Triomphe in Paris. It must be the Tour de France, no cars, just groups of cyclists (Figure 14).

Each cyclist is a carbon atom, and when they are in a group, we consider them a molecule. Here are the rules of our cycle competition. Each peloton can have four to six cyclists. The roundabout is a toll road; you must enter with four coins and pay at the exit. You cannot exit in the same round as you enter. As a result, you must pass your coins to the team so that another team member can exit. In each round, two cyclists (two carbon atoms) enter, and two cyclists exit. At the entry point, the team can only have four members, because six

---

[j]   Metabolic reactions: We call chemical reactions that occur in our body and are catalysed by enzymes metabolic reactions or metabolism, if we refer to all of them. Enzymes were called ferments when biochemistry started as a discipline.

is the maximum. Then two cyclists enter with four coins each. They pass the coins to other members of the team. Before the teams come back to the entry point, two team members exit one by one and pay the toll of four coins each. Now the team is ready to pick up another two new members, and the whole circuit is repeated.

Figure 14. The Krebs cycle in action. Cyclists are carbon atoms, and coins are electrons. For explanation, see text.

This scheme can go on forever if you have groups of two cyclists, which are nutrient fragments coming from glucose and other nutrients. The teams never exhaust, because the team members remain in the cycle only for two to three rounds before they are out again. Once you are out, you are out, because you gave away your money. Let us now reveal the parts of our cycle competition. The coins are electrons plus a proton. The cyclists with money are nutrient fragments derived from our foodstuffs, such as sugar, fat, or protein. The cyclists without money are carbon dioxide. So what does the cycle achieve? The two-cyclist groups with money (nutrient fragments) are entering. The cyclists (carbon atoms) are then stripped of their money (electrons), leaving as carbon dioxide (one by one). The money can be used to make ATP, as we will see in a moment.

Importantly, we have broken up the nutrient fragments of two cyclists into individual cyclists.

One more detail, before nutrient fragments enter, the cyclists must accelerate. For this, one carbon atom is split off as carbon dioxide, and the remaining fragment gets a push by another vitamin-derived coferment (coenzyme). This was discovered by Fritz Lipmann whom we met right at the end of the discovery of ATP. The acceleration process requires energy provided by the departure of carbon dioxide, and once the cyclists have sufficient speed, they can enter the cycle.

The cycle competition itself was worked out by Hans Krebs (1900–1981, Figure 15),[52] after whom the cycle has been named. Krebs and Lipmann received the Nobel Prize in 1953 for these discoveries. The toll paid by the cyclists upon exit was initially proposed by Heinrich Otto Wieland (1877–1957, Nobel Prize 1927 for the chemistry of bile acids) and the Swedish biochemist Thorsten Ludvig Thunberg (1873–1952). Parts of the cycle were recognised earlier by Albert Szent-Györgyi, who received the Nobel Prize in 1937, but he envisioned a more linear cycle race in the reverse direction.[53] Thorsten Thunberg (1873–1952) had proposed a cyclic reaction scheme for carbon dioxide generation in the 1920s but could not provide sufficient evidence to support it. As we have seen several times, posterity gives credit to the scientists who convinced the world about certain facts not necessarily to the people who had the initial idea.

Hans Krebs had Jewish parents who believed in assimilation and sent Hans to a protestant school. His mentor at the university was Otto Warburg, with whom he published well. This did not make a difference, and he was dismissed from his university post in 1933 and moved to Britain, where he worked out the Arc de Triomphe cycle in Sheffield.

Figure 15. Hans Krebs with wife Margaret, and Fritz Lipmann with wife at the Nobel Prize reception. Public domain, via Wikimedia Commons.

The cycle is now known as the Krebs cycle. Albert Szent-Györgyi, who worked at the University of Szeged, joined the Hungarian resistance. Although Hungary was allied with Nazi Germany, the Hungarian prime minister sent Szent-Györgyi in 1944 to Istanbul to begin secret negotiations with the allies. The plot was leaked, and Adolf Hitler himself issued an arrest warrant for Albert Szent-Györgyi. He escaped from house arrest and remained a fugitive for the rest of the war.

We saw that nutrient fragments (carbon atoms with hydrogen and oxygen), after paying the toll (removal of electrons and protons), exited the Krebs cycle as carbon dioxide. This is our contribution to global carbon dioxide emissions. But what happens to the coins? The coins are briefly stored by the 'coferment'. To be precise, it stores protons and electrons. It is derived from vitamin B_3, and we briefly mentioned its proper name before, namely NAD^k (nicotinamide adenine dinucleotide, in case you ask). Vitamins are so-called micronutrients because we need only tiny amounts for our body to function. If storing electrons and protons is so important in the breakdown of nutrients, why do we only need tiny amounts? It is because we store them only very briefly. Just imagine you use coins to pay for every coffee and meal, but you only have space for a handful of coins in your wallet. To live, you will have to change banknotes frequently into coins (break down nutrients) and use them. Let us assume you can only take as many coins as your wallet holds, then you must refill and spend at a similar rate. This analogy also tells you why our body does it like this: it prevents overspending. There is no 'Afterpay' in our mitochondria, but as we saw earlier between muscle and liver tissue, there is 'Afterpay' in the form of lactate, as discovered by Gerty and Carl Cori. What do we do with the coins: we spend them to get something done? What happens is that the electrons (coins) go through a series of reactions where they are added onto iron atoms (reducing iron) and then removed again (oxidising iron). The iron atoms are embedded in larger structures that give them different propensity to accept and donate electrons. The adding and removing causes a colour change, which led to their discovery by David Keilin (1887–1963) in 1924. He wrote, 'I must admit that this first visual perception of an intracellular respiratory process, was one of the most impressive spectacles I have witnessed in the course of my work.'[43]

We will now switch to a different analogy to better explain what happens next. This time the coins (electrons) are rolling down a couple of steps each fitted with a paddle wheel (Figure 16).

---

k   To be precise, Warburg's coferment turned out to be NADP, a difference that only biochemists want to know. Hans von Euler's coferment is NAD.

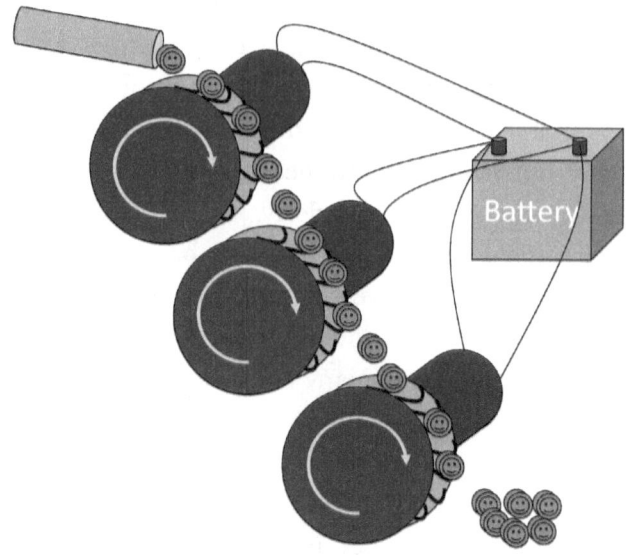

Figure 16. Power generation in mitochondria. Coins (electrons) roll down several steps, turning a paddle wheel at each drop. These are connected to generators that charge a battery.

When the coins reach the bottom, they have run out of energy, but at each step, they turn a paddle wheel, which is attached to a generator. When the coins (electrons) have run out of energy, water is produced. The electrons combine with oxygen and protons to form water (H_2O). The paddle wheels produce a current that we can use to charge a battery. The mitochondria themselves are the batteries. They are surrounded by insulating films called membranes. The membranes are most readily compared to soap bubbles. That sounds very fragile, but at the scale of a mitochondrion (one thousand of a millimetre), they are quite robust. Similarly, the lithium batteries in our mobile phones have two compartments separated by a membrane. Both compartments can store lithium ions in separate ways. When we charge the battery, lithium ions accumulate on one side and combine with electrons to settle down as lithium atoms. When discharging, the lithium atoms release the electrons, and the resulting lithium ions move to the opposite side to settle down at the other end.

The electrons move through the cable and power our mobile phones before combining with the cobalt ions on the opposite side, the buildup of charges is neutralised by the lithium ions. In our mitochondria, it is protons that are moving to the other side.

The reader will be quite familiar with protons perhaps without knowing. Just taste the juice of a lemon. Our bodies can identify protons very well. In fact, we love protons – we like pickled gherkins, we like fizzy drinks, and we like wine. Lots of protons in all of them. We do not like stuff that takes away protons, such as bicarbonate (dishwasher detergent) and caustic soda. Back to our paddle wheels. The coins were collected by the coferment upon exit from the Arc de Triomphe cycle race and delivered to the paddle wheel cascades. When the coins roll down and turn the paddle wheels, energy is released. This energy is used to push protons across the membrane and charge the mitochondria. This is the equivalent of the generators driven by the paddle wheels charging a battery. You may think, *This looks like a complex machine. How does it fit into a mitochondrion that is one thousand of a millimetre in size?*

This is one of the most fascinating aspects of modern biochemistry that we know the structure of these complex machines (often called nanomachines) down to the atom. We know where the electrons zip through and how the complex wiggles to push protons across the membrane. These machines are quite small, only a couple of millions of a millimetre long. But the energy on this scale is immense. The battery is charged to the point where arcing is not far away. Once we have charged the battery, we can finally use it to make ATP!

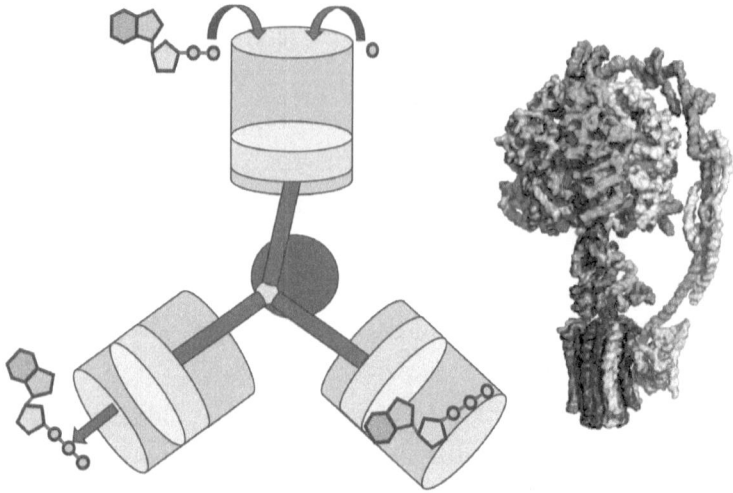

Figure 17. Left: Mechanistic analogy of the ATP synthase mechanism. The radial engine has three cylinders in different positions. Each cylinder cycles through the same positions. (1) ADP and phosphate enter the active site, (2) ADP and phosphate fused to form ATP, (3) ATP expelled. Right: An atomic scale rendering of the ATP synthase. Grey shades indicate different subunits that assemble to generate the complete nanomachine. The cap of the mushroom stands still and is held in place by the scaffold structure on the side. The stalk rotates.

How do we make ATP using a battery? For this, we have another magnificent nanomachine called the ATP synthase (Figure 17). It is driven by the proton current that the battery generates. As a result, it must sit in the same membrane that was charged and looks like a mushroom. But it is more dynamic than a mushroom. The stem of the mushroom rotates (driven by the proton current), but the umbrella-shaped cap is held in place. The cap is only the overall shape. In reality it works like a three-cylinder radial engine. The three cylinders point in three directions with each cylinder being in one of three positions: position 1 – cylinder getting loaded with ADP and phosphate, position 2 – ADP and phosphate pushed together forming ATP, and position 3 – ATP pushed out of the cylinder. The rotating stalk is asymmetric, pushing each cylinder into the three distinct positions as it rotates and pushing the ATP out once it has been made.

The radial engine analogy is fitting because we can run the engine in reverse mode using ATP to rotate the axle, but the normal biological mode is the production of ATP. Now we have finished our cellular journey starting from nutrients to the production of ATP.

But there is one important detail to add. Mitochondria were once independent bacterial cells, and the ATP is produced by their machinery within them. However, the vast majority of ATP is used in the cytosol, the main inner space of the cell. Evolution has solved this problem with the ADP/ATP translocase (translocator), which brings ADP into the mitochondria and exchanges it against ATP just produced. It is tempting to think about the translocase as a revolving door where one person (ADP) can move in while another person (ATP) moves out, but the more realistic comparison is the 'kissing gate' (Figure 18).

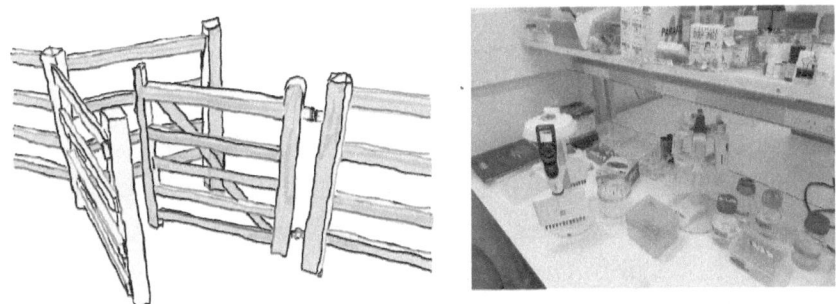

Figure 18. The kissing gate (left) as an analogy of transport across cell membranes. The automatic pipette is a symbol of research and biotechnology (right).

In the kissing gate, the gate is pushed to one side, and the person enters and moves to the corner, upon which the gate can be moved to the opposite side. The next person can then enter from the other side and move into the opposite direction. The discovery of the ADP/

ATP translocase and of its kissing gate mechanism is linked to two poisons, namely atractyloside and bongkrekic acid. Atractyloside poisoning is an infrequent form of poisoning often in the form of herbal medicine.[54] Its toxicity was soon linked to ATP production in mitochondria. It is found in many plants, particularly *Atractylis gummifera*, and can cause fatal liver and kidney damage. Bongkrekic acid is produced by the bacterium *Burkholderia gladioli* when it has time to grow on coconut and corn-base products mostly in China and Indonesia. A traditionally fermented coconut product called *tempe bongkrek* can be contaminated with the bacteria. Mortality is 40%–60% of affected individuals. Initially, there is too much glucose because it cannot be burned; later there is too little because it costs ATP to make glucose. Dizziness, low blood pressure, low body temperature, and heart problems ensue. Death can occur one to twenty hours after onset of symptoms.[55] The ADP/ATP kissing gate was discovered in 1965 by Martin Klingenberg (born 1928) and Pierre V. Vignais (1926–2006). Confirmation that atractyloside blocked the translocase was an important verification of its role in mitochondrial energy production.

At this point, I want to make a little detour into the research laboratory. Many of the experiments that revealed the basis of cellular energy generation make extensive use of automatic pipettes (Figure 18). In fact, pipettes can be found universally on photos depicting laboratory research in biotechnology. In these devices, movement of a piston attracts or displaces a small amount of air, and the liquid follows. This device was developed by the need of research laboratories working on ATP and its role in cellular functions. Heinrich Schnitger had the arduous task of analysing hundreds of small liquid samples in Theodor Bücher's laboratory in Marburg. We met Theodor Bücher briefly because he isolated one of the enzymes that make ATP in glycolysis. Heinrich Schnitger was a tinkerer and did not like to use mouth suction to remove liquids using thin glass tubes, which was the common procedure at the time. He left the laboratory for a couple of days and came back with the first piston pipette which was then improved into the workhorse of modern biotechnology.[56]

Let us summarise because it was a lot of detail: Nutrient fragments move into the mitochondria. Protons and electrons are withdrawn (the toll in the Arc de Triomphe cycle), and carbon dioxide (cyclists without money) is generated. The toll money is used to make ATP. With it, we charge a battery with protons, and the resulting current drives the ATP synthase (radial engine), which generates ATP. Lastly ATP leaves the mitochondria in exchange for ADP. Next, we will see how muscle uses ATP to contract. Before we do so, another look back into history. To work out how cellular respiration generates ATP was an enormous task, like the discovery of ATP. Again, the key researchers were decorated with Nobel Prizes.

Otto Warburg is one of the key figures who established key principles of cellular respiration. In 1908, he started his investigation using sea urchin eggs. In 1912, he discovered an enzyme which activates oxygen which he named the *Atmungsferment* or respiration enzyme. The enzyme did not appear in cell extracts made by Buchner's method but was intimately associated with cellular structure. The enzyme was inhibited by cyanide, which Claude Bernard had shown to inhibit cellular respiration. This enzyme is the last of the paddle wheels shown in Figure 16. After that, the coins have lost all of their value and can combine with oxygen and protons to make water. As we saw earlier, water is one of the end products of combustion, but now it is generated without producing much heat. Instead, we captured the energy as ATP. In 1914, Otto Warburg squashed see urchin eggs and identified small particles (mitochondria), which used up oxygen and produced carbon dioxide. David Keilin (Figure 19) showed in 1924 that proteins in yeast change colour depending on the state of respiration.[43] He observed the same proteins in many other cell types particularly in insect flight muscle cells and suggested that they were involved in cellular respiration. He also found that the *Atmungsferment* was one of the proteins that changed colour in the presence and absence of oxygen. It took many more years and researchers to identify all the paddle wheels operating inside mitochondria, and equally important how the coins rolling down the wheels or more precisely how electrons move from one iron complex to the next iron complex in the membrane of the mitochondria,[45] which causes the colour changes. Otto Warburg received the Nobel Prize in 1931 for his fundamental discoveries in cellular respiration while David Keilin missed out.

Figure 19. Pioneers of mitochondrial respiration: David Keilin (left, Wikimedia Commons) and Albert L. Lehninger (right, National Library of Medicine).

In 1949, Albert L. Lehninger (1917–1986, Figure 19) and Eugene P. Kennedy (1919–2011) demonstrated that mitochondria contained the complete machinery to make ATP associated with respiration. In 1951, Lehninger further demonstrated that for every two electrons (coins) running down paddle wheels inside mitochondria, three ATP are produced.[57] In 1959, Youssef Hatefi was instrumental in identifying the paddle wheels that harvest energy from electrons and how they combine to generate the energy for ATP production. Peter Mitchell (1920–1992, Figure 20) had the ingenious idea that mitochondria work like batteries and that the charge of the battery is used to make ATP. Peter Mitchell received the Nobel Prize in 1978.

Paul D. Boyer (1918–2018, Figure 20) received the Nobel Prize in 1997 for his insight that ATP synthesis occurs via a radial engine that rotates through three different states. John E. Walker (Figure 20) shared the Nobel Prize with Paul D. Boyer for working out the structure of the nano-radial engine called the ATP synthase.

Figure 20. Discoverers of the mitochondrial ATP synthesis mechanism. Left: Peter D. Mitchell (Image Ref. [58], Creative Commons); middle: Paul. D. Boyer with his wife (Wikimedia Commons); right: John E. Walker (Wikimedia Commons).

Peter Mitchell had a hard time convincing his fellow scientists that mitochondria worked like batteries and that their electric power was used to make ATP. Most researchers looked for a metabolite or protein with an attached phosphate that could transfer it onto ATP, similar to the process we discussed in chapter 2.[2] The debate became so heated that it was called the 'ox phos war'[59] (*oxidative phos*phorylation being the scientific term for the making of ATP during respiration). The war lasted for almost twenty years before a peace agreement was settled through a joint publication.[60] Peter Mitchell enjoyed the combative spirit of the field, writing in 1988,

> The years after the late 1950s were a particularly difficult time for me, when most biochemists (but not David Keilin) were rejecting suggestions concerning chemiosmotic coupling [the battery-driven ATP-making process proposed by Mitchell]. I had to work

very hard to get my colleagues to take these ideas seriously. Now, I find it a little sad that this work has become so much taken for granted that it is as though the chemiosmotic theory was self-evident from the beginning. (Cited after Ref. [59])

One of the researchers who was initially opposed but later convinced by Peter Mitchell's theory and helped to convince many others in the field was Efraim Racker (1913–1991).[2] Racker was taking his medical exams in Vienna when Hitler annexed Austria. He was banned from university but could just complete his exam. He then moved to the United States via Britain where he carried out his seminal research work. He was the first to isolate the ATP synthase mushroom and was later able to make a simple soap bubble with the ATP synthase and a light-driven nanomachine that pushes protons across the membrane, thereby charging the model battery. Provided with only ADP and phosphate, the tiny soap bubble produced ATP when light was shone onto it.

In Table 2, I have listed major discoveries elucidating cellular energy production associated with cellular respiration. As with glycolysis, the list is not complete, and many more researchers contributed to it. Particularly the flow of electrons (coins) down the paddle wheels is a research field in its own right, and David Keilin and Otto Warburg only started it.

Now we can move back to the role of ATP in muscle. Already in 1864, Willy Kühne[61] isolated a protein from muscle that he called myosin and considered it responsible for the rigour state of the muscle. Right at the time when ATP was discovered by Karl Lohmann, Militsa Nikolaevna Ljubimova and Alexandrovich Engelhardt[61] reported that myosin could split ATP into ADP and phosphate. Willy Kühne's method was used by Albert Szent-Györgyi, whom we met earlier, to isolate myosin and who could show that these isolated fragments contracted upon exposure to ATP.

Table 2: Major discoveries elucidating cellular respiration and energy production.

Year	Name	Discovery
1925	David Keilin	Discovery of proteins that change colour during respiration
1926	Otto Warburg	Discovery of an oxygen activating enzyme (Nobel Prize 1931)
1929	Hans von Euler-Chelpin	Isolation of a coenzyme of fermentation (Nobel Prize 1929)
1937	Albert Szent-Györgyi	Electron withdrawal during metabolism (Nobel Prize 1937)
1937	Hans Krebs	Discovery of the citric acid cycle (Krebs cycle, Nobel Prize 1953)
1937-1941	Herman Kalckar Vladimir A. Belitser	Production of ATP during respiration
1947	Fritz Lipmann	Chemical activation of metabolites required to enter the Krebs cycle (Nobel Prize 1953)
1948	Albert Claude	Mitochondria are the power plants of a cell (Nobel Prize 1974)
1949	Albert L. Lehninger Eugene P. Kennedy	Mitochondria carry out respiration and produce ATP independently
1959	Youssef Hatefi	Beginning of the identification of the complexes that harvest energy from electrons
1961	Peter Mitchell	Mechanism of mitochondrial ATP production (Nobel Prize 1978)
1965	Yasuo Kagawa Efraim Racker	Isolation of the nanomachine that synthesises ATP in mitochondria
1973	Paul D. Boyer	ATP synthase mechanism (Nobel Prize 1997)
1994	John E. Walker	ATP synthase structure (Nobel Prize 1997)

Szent-Györgyi described the contracting fibres as one of the most thrilling observations in his life. Similar observations were made by Joseph Needham at the same time during the war in 1942. While Joseph Needham could publish his results readily, Albert Szent-Györgyi had to hide from the Gestapo in Hungary. His publications were sent to the publisher at the same time but only got published in 1945.[62]

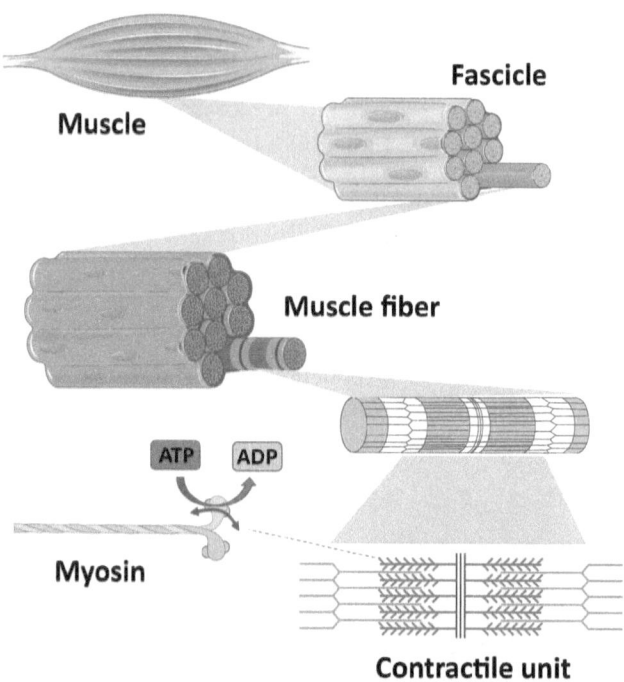

Figure 21. Muscle fibre contraction. A muscle is made up of filaments, which in turn contain myofibrils. Myofibrils have a particular pattern generated by actin and myosin molecules that can slide into each other. Myosin is an elongated molecule looking like twisted golf clubs. Isolated myofibrils contract upon exposure to ATP as first described by Albert Szent-Györgyi and Joseph Needham. Credit: Servier Medical Art.

Finally, the world could see what ATP was doing. In addition, magnesium was required to generate tension. At the same time, a second protein was identified in muscle extracts and called actin. Collectively, actin and myosin make 85% of muscle protein. Both proteins are arranged intriguingly (Figure 21), giving birth to what is called the sliding filament theory. This theory was proposed by Andrew Fielding Huxley (1917–2012) and Rolf Niedergerke (1921–2011) and independently by Hugh E. Huxley (1924–2013) and Jean Hanson (1919–1973) in 1954 and was based on light microscopy, electron microscopy, and x-ray imaging of intact muscle fibres.[63] These revealed the periodicity of the muscle fibre, which is caused by the arrangement of myosin and actin (Figure 21).

When imaging was performed in relaxed and contracted muscle, characteristic changes were observed, which were most convincingly explained by actin and myosin filaments sliding together like interdigitated fingers (Figure 21). Myosin molecules look like two golf clubs with their shafts twisted together. The heads of the golf clubs point outwards, and unlike a golf club, the heads can move back and forth. Imagine numerous of these units bundled together in a staggered way, the heads all pointing outward. Now imagine two of these bundles glued together in the centre at the ends of the shafts (Figure 21). The second protein actin forms long ropes surrounding the myosin bundles. The ropes are attached to a solid support and are long enough that one myosin bundle easily fits between with a good gap to the solid support. This is one contraction unit of a muscle fibre. Five hundred of these units fit head to tail in a millimetre of muscle fibre and about two thousand fibres in parallel bundles make a muscle (Figure 21).

As Szent-Györgyi found out, if you add ATP to these fibres, they contract; and this is achieved by sliding the thick filaments (myosin pullers) between the thin filaments (the actin ropes). In a tug of war or rope pulling, it is necessary to get a good grip on the rope and then to generate muscle tension to pull yourself along the rope (Figure 22). This is what happens in a muscle fibre as well.

The interesting aspect is that splitting of ATP to ADP does not directly pull the fibres but generates pre-tension.[1] It is like pulling the hammer of a mousetrap and arresting it. To set the mousetrap, it cannot be attached to the ropes.

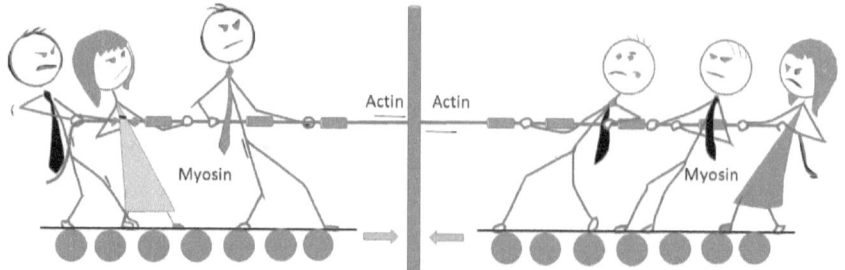

Figure 22. The sliding filament mechanism of muscle contraction. Myosin filaments pull along actin ropes.

Instead, the binding of ATP releases our closed mousetrap hammer (a.k.a. golf club head) from the ropes, allowing the trap to be opened by splitting ATP to ADP and phosphate. Once the trap is set, the hammer is allowed to bind to the ropes again. The moment it binds to the ropes, the catch is released, the hammer flips, and the power stroke occurs. This happens to many filaments at the same time, pulling them along the ropes. This cycle repeats as long as ATP is available and until the myosin bundles reach the wall to which the actin ropes are attached. We can now explain rigor mortis, the stiff muscles occurring after death. When ATP is no longer regenerated in a dead cell, it cannot release the bridge between myosin heads and actin ropes. All hands are in tight grips of the ropes, and the muscle cannot move.

When alive, we regenerate ATP all the time. Now you may wonder, why aren't our muscles always cramped? This is because there is wrapping around the ropes, so you cannot have a grip. Remove the wrapping and you can start pulling. This is where another ion is used, namely calcium. In your muscle fibres, you have tanks with a tap containing highly concentrated calcium ion solution. When you open the tap, calcium ions rush out of the tanks into the muscle fibres. The calcium binds to the wrapping, opening a gap where the myosin can have a grip. Now the pulling starts. We have postponed the problem because once the calcium flushes out, the muscles will inevitably contract. Something must pump it back into the tank. Indeed, attached to our tanks are calcium pumps or vacuum cleaners that suck up all the calcium ions, returning them to the tank. These are special vacuum cleaners that do not remove any dust, just limestone dust (a.k.a. calcium ions). Vacuum cleaners need energy, and you might have guessed it – the energy to pump calcium from the muscle fibres back into the tank is provided by ATP. Because this analogy is so fitting, these nanomachines are called calcium pumps even by biochemists. They were discovered by Werner Hasselbach (Figure 23) and M. Makinose (Figure 23) in 1961.[64] Before then it was already known that a 'relaxing factor' was present in subcellular preparations of muscle tissue. Hasselbach and Makinose showed that it was activated by ATP and removed calcium. Thus, ATP is needed both for the contraction and relaxation of muscle fibres.

These days we have again a detailed atomic view of the calcium pump and its dancing movement while it is shovelling calcium back into the stores (Figure 23). These details were worked out by Chikashi Toyoshima and Hiromi Nomura in 2002.

The final connection we must make is to the cattle fence, which I mentioned in chapter 2 as a trigger for muscle contraction. You may have touched one if you live or grew up in a rural area. This gives you an electric shock and a forceful muscle twitch. The physiologic electric shock to contract your muscle is much smaller and more local. It is generated by a local short circuit of your cellular battery. This time not in the mitochondria, because they are busy producing ATP and do not want to be interrupted, but in the membrane that surrounds our muscle fibres.

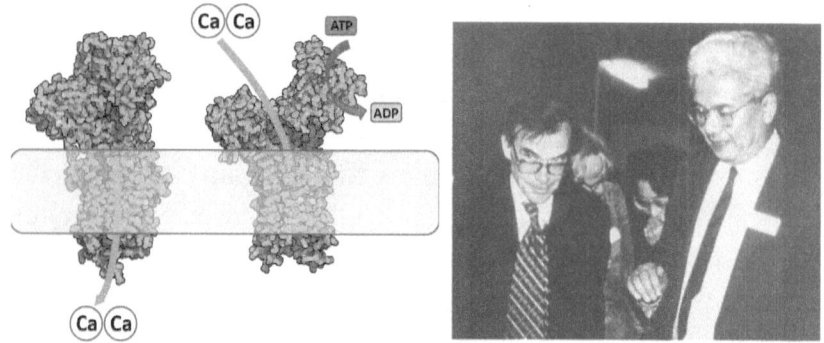

Figure 23. Left: A rendering of the structure of the calcium pump. There are dramatic changes to its shape driven by splitting ATP that allow the pump to load calcium ions on one side of the membrane and to release them on the other. In this case, the membrane encloses the calcium tanks. Right: The discoverers of the calcium pump, Werner Hesselbach (left) and M. Makinose (right), Image Ref.[58], creative commons).

As in a battery, the short circuit collapses the voltage, and this is noticed by the tanks. The taps on the calcium tanks have a voltage measuring device and open the moment the voltage drops. While it is an interesting experiment on our body physiology to touch a live cattle fence, please do not touch a live wire in your house. The voltage is far higher than in a cattle fence and instead of being unpleasant can cause a heart arrest depending on how the current flows through our body. We will talk about our hearts in the next chapter, but in essence, it is a big muscle that regularly squeezes a shot of blood out of its chamber into the circulation. We can arrest our hearts with an overpowering voltage, but we can also kick-start it with a voltage shock from a defibrillator.

Back to the short circuit that triggers muscle contraction. This is provided by nerve fibres emanating from our spine and fanning out to each muscle filament. Our muscle movement is voluntary. To initiate contraction, our brain instructs the nerve fibres to release a chemical, the neurotransmitter[1] acetylcholine, close to the muscle filaments.

---

[1]   Neurotransmitter stands short for a 'sender of nerve signals'.

We already met acetylcholine in the previous chapter because its synthesis was one of the first biochemical reactions known to require ATP. The acetylcholine binds to certain proteins, called receptors, in the membranes of the muscle filaments causing the initial short circuit. We will meet this group of proteins in detail in chapters 5 and 8. For now, we want to focus on muscle contraction.

Again, the question arises why our muscles are not constantly cramped once the chemical has been released. Our body breaks down acetylcholine very quickly, and that lets our muscles relax. There are artificial chemicals that prevent the breakdown. We can use them as pesticides, but the more advanced versions are much better known for their nefarious use as poisons. Alexei Navalny, the Russian opposition figure, was poisoned with these chemicals in 2020. As a treatment, Alexei Navalny was placed in an induced coma and put on a respirator because lung muscles are affected as well. Eventually, the poison is washed out of the body, and recovery is possible if acute life support is available.

The discovery of acetylcholine as the chemical that causes muscles to contract is an intriguing story.[65] Otto Loewi (1873–1961, Figure 24) worked at the University of Graz on the transmission of nerve activity. In 1921, he had an experimental insight during a dream and jotted down a couple of notes after waking up from the dream during the night. Unfortunately, he could not decipher his notes the next day. The dream returned the next night, and this time he made more careful notes. For the experiment, he used frog hearts. You may have noticed that frogs were the workhorses of biochemistry before the Second World War. Otto Loewi prepared two hearts, one with nerves attached, the other without. The heart has an autonomous beat, which can be maintained for a while until the ATP runs out. Both hearts were filled with salt solution, not blood. Loewi electrically stimulated one nerve for a couple of minutes. Depending on the nerve, this slows down or accelerates the autonomous heartbeat. Then he transferred the salt solution from the stimulated heart to the other heart. When he transferred the salt solution from the accelerated heart, the heart without attached nerves accelerated as well and vice versa – when the solution from the slowing heart was transferred, the other heart also slowed down.

Figure 24. Discoverers of chemical neurotransmission. Left: Otto Loewi. Right: Henry Hallett Dale (Wikimedia Commons).

The nerve that slows down the heart rate is called the vagus nerve, and Otto Loewi called the substance that must have been released into the salt solution the 'vagus stuff'. It turns out that the 'vagus stuff' was acetylcholine. The 'accelerating stuff' turned out to be noradrenaline. At the time, this was a major leap because most neuroscientists believed that electric nerve impulses were conducted by biological cables. There was no place for chemical 'stuff' in nerve communication. Loewi admitted that he would have rejected that experiment himself if he had thought about it during daytime because it was unlikely that enough 'stuff' would leak into the salt solution inside the heart.

Otto Loewi received the Nobel Prize in 1936 for the discovery of the chemical transmission of nerve impulses. Henry Dale (1875–1968, Figure 24) was the first to isolate acetylcholine from tissues and

could show that muscle contracts upon exposure to it.[66] He shared the Nobel Prize in 1936 with Otto Loewi.

Otto Loewi was already sixty-five years old when the Nazis invaded Austria. As a Jew, he was arrested at gunpoint at home and brought to prison. He was released after the intervention of several prominent international physiologists congregating at a physiological congress in Switzerland. Otto Loewi had to rescind his Nobel Prize money to the Nazis before being allowed to leave Germany for London.

Now that we have covered the main steps involved in muscle contraction, I want to summarise the important discoveries in Table 3.

Table 3: Major discoveries associated with muscle contraction.

Year	Name	Discovery
1864	Wilhelm Kühne	Isolation of myosin
1936	Otto Loewi, Henry Dale	Discovery of acetylcholine Nobel Prize 1936
1939	Vladimir Alexandrovich Engelhardt, Militsa Nikolaevna Ljubimova	Myosin splits ATP
1942	Bruno Ferenc Straub	Discovery of actin
1942	Joseph Needham, Albert Szent-Györgyi	Contraction of muscle requires ATP
1954	Andrew F. Huxley, Rolf Niedergerke; Hugh Huxley, Jean Hanson	Sliding filament theory
1961	Werner Hasselbach, Madoka Makinose	Calcium pumps

Again, we have covered a lot of detail, which is worth summarising (Figure 25). Our mental decision to lift a weight triggers the release of acetylcholine from nerve fibres close to the muscle filaments. This causes a short circuit of the membrane, which in turn opens the taps of the calcium tanks. Flooded with calcium, the wrapping around the actin ropes is removed, and the heads of the myosin golf clubs attach to the ropes, which triggers the power stroke, pulling the myosin filaments along the rope.

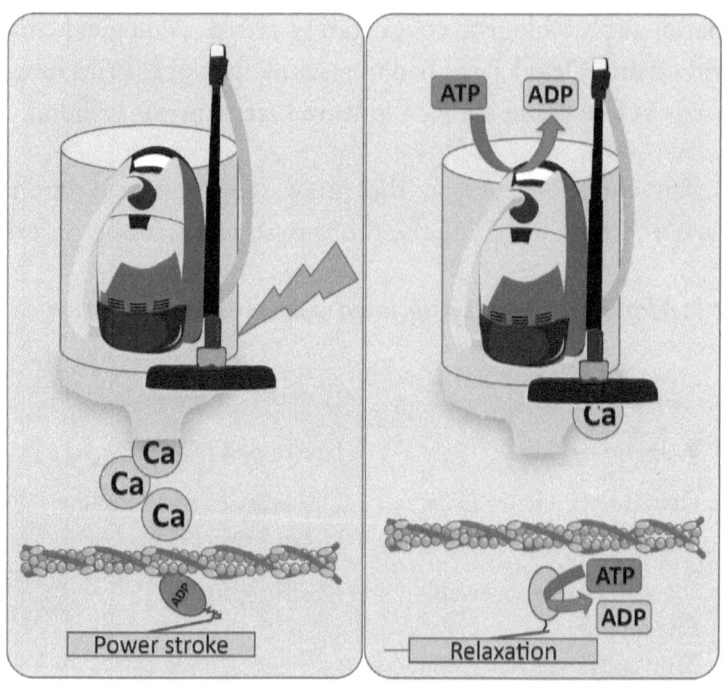

Figure 25. Muscle contraction and relaxation. An electrical shock releases calcium ions from storage tanks and initiates the power stroke. ATP releases the myosin heads for the next power stroke and spans the hammer. When the electrical shock has passed, calcium can be mopped up, and muscle relaxes.

ATP releases the golf-club heads from the rope and pulls the heads back like the hammer of a mousetrap. In the process, ATP is split into ADP and phosphate. If the wrapping is still off, the cycle continues until the calcium vacuum cleaners mop up calcium using ATP to bring it back into the tanks. Because the calcium pumps work all the time but are slower than the calcium flood coming from the tanks, our brain must tell our nerves to stop releasing acetylcholine. The remaining acetylcholine is broken down (unless we were poisoned) and our muscles relax. You can imagine how complex the orchestration must be to let us run efficiently, jump, or walk.

The groundwork is laid, and we can return to our athlete. This time we are looking at 400 m running, which takes less than a minute for a professional athlete. In our muscles there is ATP, but not a lot. It would just last for two to four seconds. In addition, we have the phosphagen system, which can rapidly recharge ADP to ATP. It is so rapid that the value of ATP never dips down a lot, just a little bit. This system is used a little longer, about thirty seconds which includes the ATP itself. Now the next store kicks in, which is glycogen. As we learned earlier, glycogen is a polymer of glucose and is readily broken down into individual units of glucose. In a 400 m sprint, blood flow is not fast enough to provide sufficient oxygen to burn the glucose to carbon dioxide and water. Instead, the sprinter uses the lactate credit card system. This provides quick energy to restore ATP from intermediates of glycolysis but comes at a cost of accumulating lactate. For 400 m, this is perfectly fine. While the athlete is running the 400 m, blood flow increases, and more oxygen is coming to the muscle cells. As a result, the burning of glucose to carbon dioxide and water occurs at an increasing rate. Unbeknownst to the athlete, her body regulates energy metabolism in her muscle. In the first 100 m, the phosphagen system is used up. During that time, lactate production increases to a peak at around 100 m before coming down again to be overtaken by respiration with no oxygen debt. From 400 m onwards, she is in a balance where enough oxygen flows to her muscle to sustain a steady pace without accumulating lactic acid. She can sustain maximum speed only for 200 m. At 800 m, her speed is already down to 70% of the maximum speed. Surprisingly, a marathon can still be run at 50% of the maximum speed although the energy provided by the different types of metabolism is quite different.[67] The phosphagen system can give us 4.5 g of ATP per kg muscle weight per second, the lactate credit card system about 2 g, and fully aerobic metabolism only 1 g. Only 0.5 g ATP per kg per second is provided by burning fat instead of sugar. The capacity of these resources is very different though. Phosphagen lets us go for 4–6 s, lactate production for 1–2 min, aerobic use of carbohydrates 1–2 h, and fat more than 6 h. The energy density provided by fat is enough to provide our tennis player from chapter 1 with sufficient energy for even the longest match. From there on food intake or energy drinks are required to get us further.

The soreness that develops over two days after an unusual bout of exercise has nothing to do with accumulation of lactate or lack of ATP.[68] Already one hour after the exercise, lactate levels have returned to normal. It is instead caused by mechanical disruption of protein structures in muscle fibres. This in turn causes immune cells to move to the sites of micro injuries where they produce hormone-like molecules that trigger pain sensation. In addition, there is local swelling caused by water being dragged into the area, which contributes to the pain.

In the previous paragraphs, we saw that anaerobic metabolic processes are sufficient to provide enough energy for short high-intensity exercise, but in the long term, the blood flow is limiting to bring enough oxygen and nutrients to the muscle to sustain optimal activity. But how should our body know that more blood flow is required to support muscle function? Again, ATP plays an important role, this time as a messenger.[69] We have learned that ATP levels do not change very much during exercise, but your muscle cells can release ATP during exercise. This is mediated by pores in the membrane the nature of which is still unknown. ATP itself is not the signal, rather adenosine, which we have met in Figure 1 of chapter 1. It is a fragment of ATP and is generated when ATP is broken down outside cells. ATP is normally kept inside cells but can leak out, possibly as a result of mechanical stress. Thus, it may be the mechanical stress on muscle during exercise that releases ATP, which is then quickly turned into adenosine. Adenosine is a powerful messenger that causes the dilation of blood vessels. This increases the blood flow into muscle, specifically the muscles that are contracting, not the ones that are resting.

There are more adaptations to exercise triggered by ATP. As we saw, the body must switch to burn fat when glycogen is running out during exercise. There is a special detector in every cell which checks the state of the ATP battery.[70] Remember that the ratio between ATP and ADP is like the value of a currency and that our muscles want to keep it mostly as ATP. We saw that we can use the phosphagen system to restore ATP, but there is another trick. If too much ADP is generated, two of those can be converted into one ATP and one AMP (adenosine with only one phosphate attached). Normally AMP is very low in cells but increases during exercise. There is an AMP detector that recognises this and activates fatty acid metabolism to help with recharging ATP. Daniel Atkinson (born 1921) was the first person to propose that cells detect the ratio of the different ATP breakdown products in 1964. Herman Kalckar, whom we met as the discoverer of ATP production during respiration, also discovered the enzyme behind AMP production in muscle in 1942.

After looking at the functional adaptations of muscle function, I would like to further mention a developmental specialisation of muscle fibres. We have seen that muscles go through different phases of energy metabolism to provide energy for sprints and marathons. However, human physiology as part of its evolution is even more adaptable. Instead of changing metabolism, why not specialising in the first place? As the reader will know, we can grow muscle mass through exercise, but we can even grow specific types of muscle. Type 1 muscle fibres are designed for a marathon. They have lots of mitochondria and lots of blood vessels. They cannot contract as fast as other muscle fibres and are therefore called slow-twitch muscle fibres. They don't have as much glycogen or phosphagen, but they are brilliant at the tour de Arc de Triomphe. In summary, slow-twitch muscle fibres prefer steady blood flow, want oxygen flowing, produce little lactate, and have a steady power output, but not extreme. Usain Bolt, on the other hand, trains to grow 'fast-twitch' or type 2 muscle fibres. They can contract very fast, have a lot of glycogen and phosphagen, but have comparatively few mitochondria. They are not great for a marathon, but they power you along for a 100 m sprint. The flexibility of our body is nothing short of amazing. A sprinter would have 25% 'slow-twitch' muscle fibres and 75% 'fast-twitch' fibres in their legs, while a marathon runner would have 75% 'slow-twitch' fibres and 25% 'fast-twitch' fibres. Thus, our body can recommission muscle fibres to different tasks. In case you wonder how this looks like, have a look at the display of your butcher. Chicken breast is pale because it has very few mitochondria. Mitochondria are brown in colour because of all the iron complexes. A chicken does not fly long distance. At best it flies onto the next tree to escape a fox, but commercially raised chickens even do not do that. While our beef does not get too much exercise either, the meat is a lot darker, because the cows walk around steadily. If you ever bought game meat, you will have noticed that it is much darker in colour due to a lot of mitochondria. Ducks fly much longer distances than chicken, and pigeons have some of the finest type 1 muscle fibres in the animal kingdom. Hans Krebs of Cirque du Arc de Triomphe fame used pigeon muscle to do much of his research work. Fly muscles are also good, but it is harder to get enough material. Crocodiles are extreme type 2 muscle animals. If you live in Australia, you can buy the meat, and it looks pale like chicken meat. They are fast sprinters, but only for 20–30 m. You can run away from a crocodile if you are not too close to it, where it could get you in one big leap. I hope you will never come into the situation where you must try this out, but it is a good finish for this chapter.

4

The Bar-Headed Goose

'I say, you do have a heart!' 'Sometimes,' he
replied, 'when I have the time.'
—Jules Verne, *Around the World in Eighty Days*

Without much thinking, we accept that blood circulates in our body. The heart pumps the blood through arteries out into all tissues. Here the arteries branch out into ever smaller vessels called capillaries, which then merge again to form veins, which return to the heart. It is the push of the heart against the resistance to the flow of blood through our capillaries that generates the blood pressure. While passing through capillaries, oxygen and nutrients are delivered to cells, and carbon dioxide and waste products including lactate are released back into the passing blood. To recharge the blood with oxygen, it is pumped to the lungs, where the vessels branch out again into capillaries, this time to take up oxygen. The small units where the oxygen exchange takes place are called alveoli where a blood vessel is only separated by an extra-thin cell from the oxygen in the lungs. The reason why people die from COVID-19 is because of an invasion of immune cells into the space between the blood vessel and the alveolar airspace. The immune cells invade because viruses have been detected that destroy cells in the alveolus. The immune cells extend and plug the space between blood vessel and airspace;

thus, oxygen can't pass readily into the blood. The blood coming out of the lungs is normally highly enriched with oxygen, but not when the alveoli are inflamed. As we have learned in the previous chapter, oxygen is key to make ATP. From the lung, the oxygenated blood returns to the heart to complete the cycle.

This cycle took a surprisingly long time to be discovered.[71] Up to the Renaissance, the consensus was that arteries and veins were two systems that contained blood and delivered it to organs and tissues. That is perhaps understandable because a cut into arteries or veins will cause blood to gush or leak out. Arteries contained air or air-enriched blood, which brought vital spirits to organs; veins contained nutrients and were thought to originate from the liver. For this idea to work, air-enriched blood coming from the lungs had to transfer between the two chambers of the heart through pores before it could go out to all tissues. Blood flow was considered a slow one-way street ending in the periphery where it somehow disappeared. Food was transformed in the liver into blood and then delivered by the veins to all parts of the body. In short, food becomes blood and blood becomes tissues. These views developed by Aelius Galenus or Galen in the second century prevailed for the next 1,400 years. The key three insights that were missing were first to recognise the heart as a pump; second to identify capillaries as the connection between arteries and veins, which had to await the discovery of the microscope; and third that blood in veins flows towards the heart, not out to tissues. The first person to challenge Galen's views was Ibn al-Nafis from Damascus who recognised a connection between arteries and veins in the lung and discovered the pulmonary circulation in the thirteenth century, but his writings remained largely unrecognised or ignored. The advent of the Renaissance slowly generated knowledge that opposed Galen's views about blood flow. Leonardo da Vinci (1452–1519) recognised the heart as a muscle. Realdo Colombo (1515–1559) rediscovered the pulmonary circulation in the sixteenth century but still thought that most venous blood flows into the periphery and only a small portion was going to the lungs to deliver nutrients. Girolamo Fabrici (1537–1619) discovered venous valves in 1574 but failed to recognise their role in directing the blood flow back to the heart.

William Harvey (1578–1657, Figure 26), who studied anatomy under Fabrici in Padua before returning to England, finally challenged Galen's doctrines. He linked the pulse with the heartbeat and recognised that the same volume of blood was generated by bleeding an animal from an artery or a vein. He also wondered why the heart has two ventricles if the blood flow to the lung was only used to provide that organ with nutrients.

Figure 26. Discoverers of blood circulation: William Harvey (left, Wikimedia commons) and Marcello Malpighi (right, Wikimedia Commons).

Moreover, he couldn't find any pores in the septum that separates the chambers of the heart. As a result, he reinstated the pulmonary circulation and recognised the heart as a pumping muscle. Once he recognised the heart as a pump, simple calculations of the ejection volume made it apparent that blood must perform a circuit. He made further experiments in animals, clamping veins before the heart and arteries after the heart, which resulted in empty pale hearts or distended purple hearts. Similar experiments with ligatures around the arms in humans confirmed his observations. Like any breakthrough, Harvey's discovery, published in 1628, was met with scepticism and rejection, which only waned after the discovery of blood capillaries by Marcello Malpighi (1628–1694, Figure 26) in 1661. Richard Lower (1631–1691), a pioneer of blood transfusions, showed that the colour change in blood from brownish blue in the veins to red in the arteries occurred as it passed through the lungs[3] and became enriched with oxygen.

As we just re-established, our body has a special circuit to pick up oxygen. We also have a circuit to pick up nutrients that is less well known, the so-called splanchnic bed. This is the part of the circulation that provides arterial blood to the intestines, stomach, pancreas, spleen, and liver. Once it has become venous blood, the vessels that pass through the intestine contain absorbed nutrients. Eventually the blood from our digestive systems is collected and enters the liver through the portal vein. This is important because our liver protects our body against toxic compounds, particularly ammonia that is generated by our intestinal microflora. We will cover this in more detail in chapter 8. Now back to the heart.

The heart uses 6 kg of ATP per day to beat about 100,000 times and to pump the equivalent of 10 tons of blood. It turns over its ATP content every twelve seconds. Of its ATP, 60%–70% is used for mechanical pumping. The remainder is used for the calcium pumps, which we met in the previous chapter. The calcium pumps constantly mop up calcium ions, which gush into the cell to initiate each power stroke of the heart. Essentially our heart is just a sophisticated autonomous muscle. Systolic contraction is initiated by calcium gushing into heart cells; diastolic relaxation is mediated by ATP-dependent removal of calcium.[72] During the rhythmic action, other ions move as well; and their concentration is restored by another ATP-driven pump, the sodium pump, which we will meet in more detail in the next chapter. Cardiac glycosides, derived from foxglove plants, are used to treat congestive heart failure, atrial fibrillation, and atrial flutter. In higher doses, these drugs will kill us because ion concentrations cannot be maintained; but in lower concentrations, they indirectly increase the amount of calcium in the heart cells. As a result, contraction becomes stronger and more regular.

Heart cells are packed with mitochondria and contain actin and myosin filaments, which we encountered as contractile fibres in muscle cells. Pacemaker cells generate the intrinsic electrical activity of the heart. At rest they have a slow beat, but this can be altered through stimulation of the sympathetic nervous system, which then releases noradrenaline and increasing the heart pace. Some of the noradrenaline leaks into the heart chamber, allowing Otto Loewi to discover the neurotransmitter noradrenaline and acetylcholine, as outlined in the previous chapter.

Unlike skeletal muscle, which we can control, we cannot order our heart to stand still. At rest the heart uses only 25% of its maximal capacity, but with exercise this can increase to >80% of its capacity.[73] Our heart has an unusual food preference. It generates most of its energy from fatty acids and smaller amounts from glucose and lactate. The heart is not a sprinter muscle. It generates 95% of its energy from mitochondrial respiration. As a result, it cannot easily adapt to reduced blood flow, which gets noticed as angina pectoris or in the worst case as a heart attack. Angina pectoris is felt as heaviness on the chest, and an acute attack can be quite painful, spreading from the chest to the left arm and lower jaw. It is an indicator that arteries that nurture the heart have become too narrow due to atherosclerosis. A short-term fix is administration of nitroglycerol. It is the same chemical that is used to blow up rocks. When it is in our bloodstream, it decomposes slowly, generating nitric oxide. This compound is also made by our own cells, but very localised to dilate blood vessels. Nitric oxide binds to a specific site in our cells; and this causes a molecule that looks very similar to ATP, called GTP, to react with itself between the phosphates and the sugar. The compound is then called cyclic GMP and causes muscle-like structures in the walls of blood vessels to relax, which causes dilation and an increased blood flow. If nitroglycerol is taken during an angina pectoris attack, the blood flow to the heart increases, causing the pain to subside. Used as a spray, it takes only seconds to work; but it is a typical pharmaceutical approach, treating the symptoms but not the cause. Stents or a bypass surgery is the next step to fix the problem more permanently. Male readers can experience the action of cyclic GMP when they have an erection. The blood flow into the penis increases, causing it to become extended and hard.

We have seen that the heart pumps blood into the pulmonary circulation. In the alveoli, oxygen can diffuse from the airspace of the lung and bind to haemoglobin enclosed in red blood cells. Haemoglobin is the truck that carries oxygen to all cells and tissues. Our red blood cells are mostly a bag full of haemoglobin. We all know the dark-brown colour of venous blood, while blood in our arteries looks brightly red. Fresh menstrual blood, as female readers will know, is arterial and bright red in colour. Binding of oxygen to the iron atoms in haemoglobin changes its colour. This is similar to the colour changes observed by David Keilin when iron complexes inside our mitochondria accept or donate electrons.

At sea level, air contains 21% oxygen, which amounts to a pressure of 160 mm Hg. Deep down in our lungs where the capillaries are, about 100 mm Hg of oxygen pressure is left. This oxygen pressure loads all haemoglobin molecules with oxygen. We can transport about 20 ml of oxygen per 100 ml of blood.[74] Because tissues use oxygen to make ATP, the oxygen pressure is much lower, causing haemoglobin to unload oxygen. Venous blood has an oxygen pressure of only 40 mm Hg, which causes haemoglobin to lose about a quarter of its oxygen. This looks like a small fraction or a massive reserve that remains in our blood, but the numbers reflect sea level. We can walk to the top of Mount Kilimanjaro where the oxygen pressure is half of that at sea level and still survive. However, many people get high altitude sickness, because it becomes difficult for the brain to extract enough oxygen from haemoglobin. Eventually, this causes swelling of the brain, which can become life-threatening. Given time to adapt to high altitude, haemoglobin has some tricks up its sleeve to optimise oxygen binding and release. First it has four sites to bind oxygen, which influence each other. This allows haemoglobin to flip between a state which is optimal for oxygen binding in the lung and another state which is optimal for release of oxygen in the tissues. The flipping is fostered by the release of carbon dioxide from the tissues. It is also fostered by the acidification of the blood when carbon dioxide reacts with water to form carbonic acid. In addition, our body can produce molecules that improve the unloading of oxygen from haemoglobin and the concentration of these increases when we adapt to high altitude. Moreover, our body produces more red blood cells when adapting to high altitude. All of this is very beneficial for sports performance, and as a result, high altitude training is used before sporting competitions. Populations who live at high altitudes permanently have additional adaptations such as improved blood flow to the brain and larger lungs.[75]

You may wonder why we are so well adapted to performing at high altitudes, given that most people live close to sea level. However, foetal blood has low oxygen pressure, and the foetus must develop for 9 months in a low oxygen environment where the pressure is only 30 mm Hg compared to the 100 mm Hg in the lungs. Foetuses have a special haemoglobin that is optimised for performance in this environment. Moreover, our cells can switch specific genes on and off depending on oxygen pressure. William G. Kaelin, Peter J. Ratcliffe, and Gregg L. Semenza received the Nobel Prize in 2019 for deciphering the mechanisms on how cells sense and adapt to oxygen availability.

Animals have developed more tricks to optimise the use of oxygen. The bar-headed goose can migrate across the Himalayas and perform admirably at an altitude of 9,000 m.[76] With little time to adapt, the goose uses large air sacs to increase the time the air is available to exchange its oxygen with the blood. Weddel seals, which can dive comfortably for 30 min, reduce the heart rate to half of the normal rate when they dive for more than 15 min. Up to 15 min no excessive lactate is produced, but it rapidly increases after that time, suggesting a switch to anaerobic exercise like a sprinter. The long delay of lactate formation is possible because the seals have large amounts of myoglobin in their muscles. Myoglobin looks like a quarter haemoglobin and can also bind oxygen. It acts as an oxygen sponge and allows the seal to maintain oxygen levels in muscle although blood flow to the muscles is reduced.[77] Terrestric animals including humans have lower amounts of myoglobin, and the effects of myoglobin on performance are more difficult to show.

Getting insight into the structure of myoglobin and haemoglobin was one of the heroic tasks of biochemistry in the twentieth century. Both were attractive targets as large amounts of both proteins could be easily isolated, and their biochemical importance was immediately obvious. Max Perutz (1914–2002, Figure 27) was raised and educated in Vienna but decided to move to Cambridge in 1936 to learn protein crystallography.[78] At the time, it was already established that it was possible to derive the structure of a small molecule from the crystals of those molecules. Many small molecules spontaneously form crystals when water evaporates from highly concentrated solutions. Table salt, for example, forms crystals out of evaporating seawater. When placed in an x-ray beam, crystals produce a pattern of diffraction spots that can be visualised on film. Mathematical analysis of the spot pattern can reveal the atomic structure of the molecule. This is relatively simple for table salt, but proteins are much larger and more fragile molecules. John Desmond Bernal and Dorothy Crowfoot demonstrated in 1934 that proteins could be crystallised for structural studies.

Figure 27. Max Perutz (left) and John Kendrew (right) who elucidated the structure of haemoglobin and myoglobin, respectively (Wikimedia Commons).

In contrast to table salt crystals, proteins had to remain in the solution from which they were crystallised. Haemoglobin was relatively easy to crystallise, a result already achieved at the beginning of the century. Max Perutz' work on haemoglobin was interrupted by the war when he was deported to Canada because of his Austrian citizenship. Through the help of many eminent scientists, he was released and transferred back to England in 1941 where he performed research to support the war effort.[78] After the war, he resumed his work on haemoglobin and was joined by John Kendrew (1917–1997, Figure 27) who tried his luck with myoglobin.

The crystals were less of a problem than the mathematics. A first insight into the structure of haemoglobin came from the American chemist Linus Pauling (1901–1994). Through theoretical reasoning, he deduced that sections of a protein could form a corkscrew-like structure called an α-helix. Among many other achievements, he received the Nobel Prize in chemistry for this discovery in 1954. This corkscrew-like structure should produce a particular diffraction pattern on the x-ray films derived from haemoglobin crystals and was detected in due course by Max Perutz in 1951. In 1953, another breakthrough occurred after the discovery that the mathematics could be solved when haemoglobin crystals were grown in the presence of

STEFAN BRÖER

mercury,[78] which causes changes to its refraction pattern. By 1959, the resolution had improved to the point that the shape of the molecule became apparent, but the models still looked more like a curled sausage than a detailed protein. John Kendrew's myoglobin made faster progress because it was only quarter the size of haemoglobin, and an atomic model was generated in the same year.

In 1962, Max Perutz and John Kendrew were awarded the Nobel Prize for chemistry for their work on the structure of proteins. It took another six years to generate a high-resolution structure of haemoglobin where all details could be appreciated. In 1970, Max Perutz finally saw the subtle difference between oxygen-free and oxygen-bound haemoglobin. Breathing at the molecular level could be visualised for the first time. Haemoglobin breathing has been likened to a playground seesaw, where one end is fixed with a spring and the other is balanced with a weight. Oxygen binding pulls on the spring and lets the other end rise.

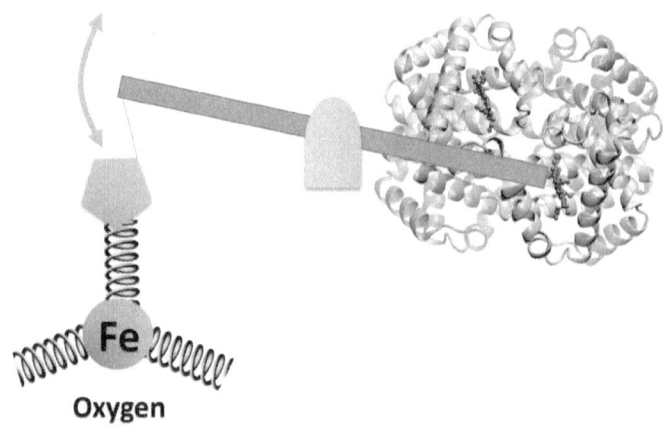

Figure 28.: Cartoon of the changes to haemoglobin upon oxygen binding. Oxygen (O=O) binds where Haem-iron (Fe) is located, pulling on the springs.

If four seesaws are tied together, one can imagine how binding of oxygen to one seesaw can affect the state of the other seesaws.[78] This mechanism underlies the switch between the haemoglobin shapes that are optimised for oxygen binding and release, respectively.

We have now seen in some detail how oxygen is transported from lungs to tissues where it is used to oxidise nutrients forming carbon dioxide and water. At rest we produce about 0.2 litres of carbon dioxide per minute. The rapid formation of carbonic acid from carbon dioxide is important to transport carbon dioxide from the place of formation in the tissues to the lung where it is converted back from carbonic acid to carbon dioxide and water. About 75% of the carbon dioxide is transported by red blood cells; the remainder is found in the blood plasma. Inside red blood cells, enzymes quickly convert carbon dioxide to carbonic acid. This releases acid (protons) which attaches to haemoglobin, thereby changing its shape, helping to unload oxygen from haemoglobin. Carbon dioxide can also directly combine with haemoglobin, further helping the removal of carbon dioxide from tissues. In the lungs, the opposite happens. Binding of oxygen releases the acid (protons), which in turn combines with bicarbonate forming water and carbon dioxide. This effect is easily observed with soda water. Add a bit of citrus juice and carbon dioxide is bubbling out more forcefully. In venous blood the equivalent of 52 ml of carbon dioxide is found, while in arterial blood 48 ml remain. We can only hold breath for a brief period because carbon dioxide will accumulate, causing haemoglobin to unload more of its oxygen, but the process will soon stop, and the brain runs out of oxygen.[79] The increased release of oxygen allows apnoea divers to ignore the build-up of carbon dioxide until the lack of oxygen becomes too severe, forcing them to breathe before they become unconscious.

A catastrophic failure of oxygen supply occurs during heart attack. This happens when one of the arteries that supply the heart with oxygen is blocked by a blood clot. As mentioned above, the heart relies on mitochondrial respiration to generate its ATP. When oxygen and nutrients get low due to the blockade of an artery, the heart tries to become a sprinter muscle and to use its small reserves of glycogen to generate ATP in the absence of oxygen. As we know from the sprinter, this cannot go forever, and even worse the lactate that builds up is not flushed away by the blood flow. After a couple of minutes, the heart cells struggle to generate enough ATP for their muscle work because of acidification caused by lactic acid. The heart cell tries to pump the protons (acid) out of the cell, but this costs energy and eventually causes more calcium to flow into the cell. This would trigger constant contraction and requires more energy to remove the calcium as well. Eventually all the remaining energy is used for cell survival instead of mechanical pumping.[80] People who experience a heart attack feel pressure on the chest; feel pain in the shoulder, arm, and neck; and become nauseous.

In the previous chapter, we learned that mitochondria perform like little batteries to generate ATP. When heart cells cannot generate enough ATP to remove the calcium any longer, they start to commit suicide.[81] Elevated calcium short-circuits the mitochondrial battery via a conglomeration of proteins in the membrane which form a pore. The pore short-circuits the battery, which cannot maintain its voltage. Calcium and water flow into the mitochondria which causes them to burst. Bursting mitochondria are unwelcome news for cells. It triggers a program called apoptosis or programmed cell death, which wrecks and dismantles the cell like demolishing an old house. It is the appearance of a specific mitochondrial protein in the main part of the cell that is the key indicator for cells to commit suicide. There are special cells that can remove the rubble remaining after cell death. However, the dead cells cannot be replaced because there are no stem cells in the heart that could develop into mature heart muscle cells. Thus, the damage is irreversible. This is the reason it is so important to remove the blood clot within two hours after a heart attack to minimise cell death. It is quite remarkable that arteries that are hardly open due to atherosclerosis still provide enough blood flow for the heart to work. However, episodes of angina pectoris are the tell-tale sign that blood flow is suboptimal for higher performance.

Surprisingly, reopening the arteries and restoration of blood flow can do even more damage. The rushing in of oxygen can damage already compromised mitochondria, pushing them over the limit. Oxygen is a very reactive molecule, and as we learned in chapter 3, there are lots of electrons on the move inside our mitochondrial batteries. These electrons can combine with oxygen, forming molecules like bleach and hydrogen peroxide. It does not take too much imagination that a flood of these molecules can further damage mitochondria that are already compromised. The good news is that we can infuse drugs when the blood flow is restored, which help compromised mitochondria to survive the onslaught of oxygen.

Provided that oxygen is available, our heart is a remarkable machine. In the up to 100 years of a human life, it beats about 4 billion times without any interruption. This is even more amazing because oxygen continuously forms hydrogen peroxide and related molecules inside mitochondria in smaller amounts. These are quenched all the time, and any smaller damage is constantly repaired by exchanging proteins and lipids to maintain function. Like in a modern airplane, all parts are regularly exchanged to maintain function, and only at the end of a 100-yearlong life this task may become too difficult to maintain with too much debris and non-functional parts accumulating. The reader won't be surprised to hear that the recycling process requires ATP to dismantle old protein and to recover its parts for future use.

5

ATP Meets Frankenstein

*Reality provides us with facts so romantic that
imagination itself could add nothing to them.*

—Jules Verne

We associate our nervous system more with energy than any other organ. It generates electricity, it communicates by electrical signals, and it responds to electrical signals. As we will see, ATP is at the heart of generating electricity in the nervous system, and it requires a large amount of nutrients to provide the energy to maintain the electricity in our brains.

When Mary Shelley created Frankenstein in 1818, she let her hero 'infuse a spark of being into the lifeless thing'. The lightning strike is a Hollywood invention but fitting in spirit. Not long before the novel was published, Luigi Galvani (1737–1798) had discovered that a frog leg muscle when pierced by a copper hook sometimes contracted when the copper came into contact with iron, most likely when some moisture was at the contact site, generating a small current. The next step was the deliberate application of a small electric shock after which the leg contracted as well. Galvani and Alessandro Volta (1745–1827) worked with electric fish to understand the connection between life and electricity. Hermann von Helmholtz (1821–1894) recognised that nerve cells generate electricity to produce messages. He also recognised that the flow of electricity along a nerve fibre was much slower than along a copper wire.

In contrast to our hearts, our brains are very selective when it comes to food choices. Under normal physiological conditions, our brain only consumes glucose and nothing else. Constant blood flow to the brain is essential to maintain consciousness, which is lost within fifteen seconds when blood flow stops, for instance because of cardiac arrest. This shows that our brain is very hungry. Indeed, it consumes 20% of our total energy requirements despite only weighing 2% of our body weight. To understand why the brain needs so much energy, we must have a brief look at the cellular anatomy of the brain (Figure 29).

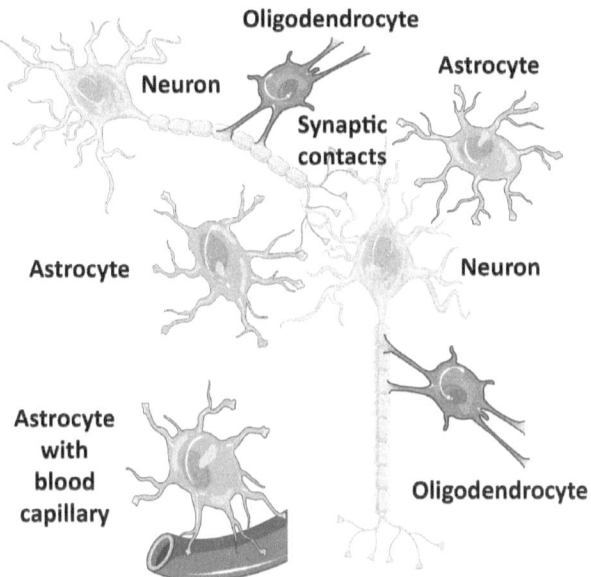

Figure 29. Cell types of the brain. Neurons propagate signals, while astrocytes support neuronal function and regulate blood flow. Contacts between neurons are called synapses; nerve fibres are called axons which end in a button-like structure, the bouton. The buttons are close to spikes on branches of the neuronal cell body called dendrites. Oligodendrocytes isolate nerve fibres. Image credit: Servier Medical Art.

The brain contains an estimated 88 billion neurons each of which has thousands of endings, which form contacts to an estimated one thousand other neurons called synapses. Neurons are not the most frequently occurring cell type in the brain. This honour goes to astrocytes of which there are about five times as many as neurons. Astrocytes are important to optimise brain function and regulate blood flow. Einstein's brain, when anatomically studied after his death, was not unusual apart from having a larger number of astrocytes.[82] There are more types of non-neuronal cells, namely oligodendrocytes, which wrap insulation around nerve fibres and microglial cells, which form the brain's immune system. Ependymal cells line liquid-filled spaces in our brain called ventricles. Lastly endothelial cells line the blood vessels that provide nutrients to the brain. In the rest of our body, there are gaps between endothelial cells allowing nutrients to seep through into the tissue. The brain instead is protected by tightly packed endothelial cells forming the blood-brain barrier. This is necessary because our brain uses common nutrients to communicate.

Santiago Ramón y Cajal (1852–1934) was the person who visualised the nervous system for the world and the first to understand its cellular structure and the resulting physiological function with remarkable clarity. Ramón y Cajal had artistic talent and used a silver stain developed by Camillo Golgi (1843–1926) to visualise the tree-like structure of neurons (Figure 30). The stain has the unintended property that it penetrates only one in one hundred neurons, which clarifies the anatomic structure as he would otherwise only see a forest packed with trees. Cajal also used foetal brain tissue which has less connections between neurons than adult tissue. For comparison, imagine a young forest where you cut out ninety-nine out of one hundred trees with a dense old forest as a landscape to trace the shape of a tree. In the absence of any experiments, he proposed that there were small gaps between neurons and not continuous cables as Golgi believed. Moreover, he postulated that there is a receiving end of a neuron, the dendrite, and a messaging side of a neuron, the axon. He recognised the existence of circuits made up by successive neurons that propagate a signal. Camillo

Golgi and Santiago Ramón y Cajal received the Nobel Prize in 1906 although both had incompatible views about the cellular structure of the nervous system. Apart from the stain, which stood the test of time, posterity confirmed all of Cajal's proposals.

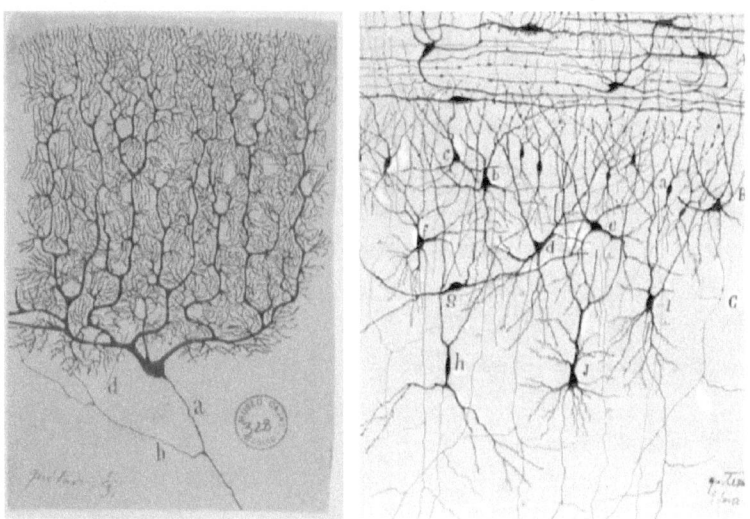

Figure 30. Two of Cajal's drawings of the structure of the nervous system. The drawings are based on slices of the brain, which were stained with a special silver solution (Image credit: Cajal Institute).

Neurons communicate with each other through nerve impulses and neurotransmitters. In the previous chapter, we encountered the neurotransmitters acetylcholine, which slowed the heart down, and noradrenaline, which increased heartbeat frequency. These neurotransmitters are also used inside our brains, but there are more than one hundred different ones for a wide variety of tasks. Nerve impulses run along axons as electrical signals in the form of a dip in voltage. In contrast to a power company, which wants to keep the voltage of the grid constant, our neurons modulate it all the time. A nerve impulse runs along the nerve fibre at a speed of 100–200 km/h. The nerve fibre of a single neuron is called an axon. Between brain areas many of them are bundled together, forming the white matter of our brain. The nerve impulse is generated by a brief short circuit of the electricity supply. In our household that triggers a fuse to blow, upon which the voltage is re-established through the supply to our houses. A similar mechanism occurs in our nerve fibres. As in mitochondria, which we met earlier, the membrane of a neuron is like a battery and generates a certain voltage. Not as high as in mitochondria but still very substantial. The first person to measure this voltage was Emil du Bois-Reymond (1818–1896). Once the battery is charged, it does not take much energy to keep it charged, but there are frequently occurring short circuits everywhere in neurons when they communicate with each other. The short circuit in a nerve fibre is brief because it is automatically terminated after a couple of milliseconds. Another nanomachine is involved in this, called an ion channel. To explain, I will use a cattle analogy (Figure 31).

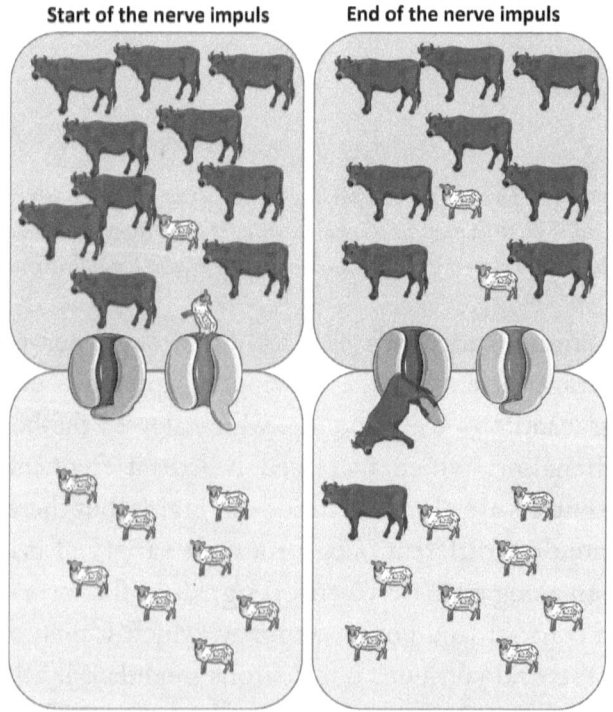

Figure 31. Analogy of a nerve impulse or an action potential. Two paddocks are connected by two gates, one for sheep and one for cows. Initially the gate for sheep opens, and some move across because the 'grass is greener on the other side'. Then the cow gate opens, and some cows move in the opposite direction to balance the numbers.

Sheep and cows passing through the gates are the equivalent of ions moving through a channel. To understand a nerve impulse, let us assume that there are two different gates separating a sheep and a cow paddock. In Figure 31, the one gate lets sheep through the other gate cows. Now we further assume that the upper paddock is for cows (the axon) and the lower paddock for sheep (the space outside cells), and both are filled with the same number of animals. Overall, we want to keep the same number of cattle in both paddocks. Cows and sheep are social animals and want to stay with each other but also find that the 'grass is greener on the other side'. These two opposing desires limit the movement of sheep and cows [m].

Initially the door for sheep opens, and because the grass looks greener, some will move across, but the majority will remain with the herd because they are social. Next, we open the gate for cows, and some will start moving in the opposite direction, but most will remain with the herd. Now the first gate closes to limit further movement of sheep. Once the same number of cows has moved in the opposite direction, the second gate closes as well. In electrical terms, the opening of the first gate generates a short circuit (imbalance of animal numbers) but then closes spontaneously. The opening of the second gate re-establishes the balance of the total number of cattle in each area (re-establishing the standard voltage). The first gate constantly monitors the voltage along the nerve fibre. When the voltage drops in the neighbourhood, the gate opens itself, and ions (sheep) rush through, which causes a drop of the voltage at the site of the channel. This is sensed by neighbouring gates which open as well, thus propagating the nerve impulse. Then the first gate shuts, and the cattle balance (voltage) is restored by opening the second door. The first door remains shut for a while, so the short circuit cannot be repeated immediately. This also ensures that the nerve impulse travels only in one direction and cannot come back. The two

---

[m]   The two opposing forces for an ion are the concentration gradient and the electrical force building up when ions follow their concentration gradient without an accompanying counterion.

opening events re-establish the total number of cattle (ions) in both areas but change the distribution of sheep and cows (ions) along the nerve fibre. Some sheep were left in the cow paddock, while cows remained in the sheep paddock. We will see in a moment how this is restored, but first a little history.

The first person to record a propagating nerve impulse, or action potential as it is called, was Edgar Douglas Adrian (1889–1977, Figure 32). He used a thin metal wire to touch the surface of a nerve fibre and recorded the electrical changes with an ink writer. He recognised that all nerve impulses look the same and that information in the brain was encoded by how frequently nerve impulses were triggered. He received the Nobel Prize in 1932. This travelling of nerve impulses, particularly along large bundles of nerve fibres, cannot only be detected by metal wires but even outside the brain by recording an electroencephalogram (EEG). The electroencephalogram was developed by German psychologist Hans Berger (1873–1941) who had an intense interest in the energy requirements of the psyche.[83]

Figure 32. Pioneers of modern neuroscience: Edgar Douglas Adrian (left, Wikimedia Commons) and Charles Scott Sherrington (right, Wikimedia Commons).

Without knowing anything about ion channels, Alan Hodgkin and Andrew Huxley (Figure 33) worked out the electrical properties of axons in 1952 at the University of Cambridge and received the Nobel Prize for this work in 1963. It is the same Andrew Huxley who also came up with the sliding filament theory of muscle movement, which proposed that muscle filaments slide into each other like interdigitating fingers. Alan Hodgkin did all this seminal work during his dissertation. At the time, axons in higher animals were way too small to be investigated with electrodes. Hodgkin and Huxley used the squid giant axon to perform their experiments, but the results have stood the test of time also in other animals including humans. Erwin Neher and Bert Sakmann massively improved the techniques in the late 1970s and early '80s to measure very small currents while working in Göttingen. They were able to record the miniscule currents of individual ion channels and were awarded the Nobel Prize in 1991. It took a long time to elucidate the structure of ion channels. Roderick MacKinnon (Figure 33) achieved this breakthrough in 1998 and received the Nobel Prize in 2003.

Figure 33. Neuroscientists that investigated the communication between neurons. From left to right: Andrew Fielding Huxley, Alan Lloyd Hodgkin, and Roderick MacKinnon (Wikimedia Commons).

We still need to answer two questions: first, what events cause the opening of the gates for sheep (drop in the voltage) in the first place to initiate the nerve impulse, and second, how are cows and sheep returned to their original location to restore the starting conditions. Inspection of Figure 29 suggests that one end of a neuron has lots of processes. There are so-called spikes on these processes, and opposite each spike you would find the end of an axon from another neuron. The end of each axon forms a button-like structure called a bouton. There is a small cleft between the button and the spike. The whole assembly is called a synapse, and it is here where the nerve impulse is passed from one neuron to the next by chemical substances which are called neurotransmitters (Figure 34).

The term *synapse* was coined by Charles Sherrington (1857–1952, Figure 32) in 1897. He wrote, 'We are led to think that the tip of a twig of the [nerve fibre] arborescence is not continuous with but merely in contact with the substance of the dendrite or cell body on which it impinges. Such a special connection of one nerve cell with another might be called "synapsis".'[84] Sherrington also recognised that there are two types of neurons: one that propagates nerve impulses, called excitatory, and one that inhibits excitatory neurons from propagating the signal to the next neuron. These are called inhibitory. He recognised that when an arm or leg muscle contracts on one side, the muscle on the opposite must relax, which is mediated by inhibitory action on the excitatory nerves of the opposing muscle. He received the Nobel Prize for these discoveries in 1932 together with Edgar Douglas Adrian.

We take the synapse plus neurotransmitter model of nerve transmission for granted today, but initially most scientists thought that electric transmission of nerve impulses was mediated by cable-like connections between cells. This is not surprising because this is how electricity is propagated by engineers, and axons look a lot like cables. The long argument between supporters of chemical transmission and cable transmission has been dubbed the war of the soups and the sparks.[65] Soups refer to extracts of brain tissue containing all its chemicals, while sparks refer to the electrical currents.

The first evidence for chemical transmission came from the experiments carried out by Otto Loewi and Henry Dale, which I described in chapter 4. The vagus nerve released a substance that caused the heart to slow down. However, these observations were not considered an argument that the same occurred in the brain. The neurophysiologist John Eccles, for instance, wrote in 1936 that the evidence for chemical neurotransmitters in the brain was almost negligible. This opened the door for biochemists and pharmacologists who instead of using electrodes (sparks) preferred to grind tissue and prepare a 'soup' for their investigations. The spark proponents admitted that acetylcholine could modulate heart pace but that the fast response of skeletal muscle to nerve impulses required direct electrical contact. Walter Cannon (1871–1945), an influential American physiologist, however, pointed out that there was a small delay of nerve transmission at the junction between nerve and muscle that could not be explained by an electrical connection. Another piece of evidence came from the plant poison curare that is applied to arrow tips by Indians of South America to quickly paralyse birds and small animals. Curare blocks the response of muscle to an incoming nerve impulse. Claude Bernard, whom we met as the discoverer of glycogen, found that although the nerve impulse could not jump from nerve to curare-treated muscle, the muscle could still contract when an electrical stimulus was directly applied to the muscle tissue.[84] We know today that curare attaches to the same protein as acetylcholine, blocking its action. Such a protein is called a receptor. It is a bit like a receptionist. If you want to deliver a package or message, you announce yourself to the receptionist. The receptionist then calls the receiving person or goes into action herself.

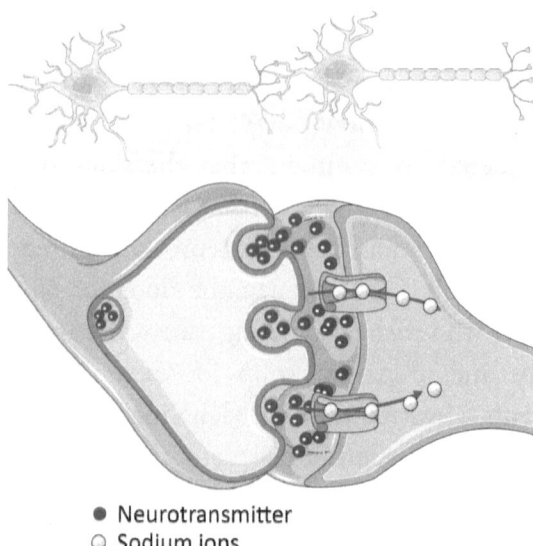

● Neurotransmitter
○ Sodium ions

Figure 34. Synapses connect neurons. An incoming nerve impulse triggers the release of packages of neurotransmitters. These bind to receptors (receptionist) on the receiving end of the next neuron. The receptor opens a pore, and ions rush in, which destabilises the voltage of the next neurons. (Servier Medical Art).

Figure 34 shows a junction between two neurons or between a neuron and a muscle cell. Note that the first neuron (left) contains packages of neurotransmitters. When a nerve impulse comes in, one or several packages are released. That floods the gap between both neurons with the neurotransmitter molecules (dark spheres). On the opposite side, they bind to receptors and announce their presence. This opens a door in the receptor and ions (light spheres or sheep, if the reader prefers) are rushing into the cell behind the synapse. This causes a short circuit, and the entire process starts again. The process has been nicely shorthanded into 'ping-squirt-ping'. It is not quite as simple because the short circuit at a single synapse is not enough to drop the voltage across a whole neuron to initiate the next nerve impulse. Remember that there are thousands of synapses attached to the branches of a neuron, called spikes on dendrites. This is where the computational part of our brain sets in because there are not only receptors that cause the voltage to drop but also receptors that increase the power output of the local power station to stabilise the voltage. Each synapse has a preferred receptor and a preferred neurotransmitter. As a result, we have synapses that stabilise the voltage and inhibit the propagation of the voltage drop and excitatory synapses that drop the voltage. The integration of all signals is like regulating the voltage in a power grid. When the use of electricity is heavy, the supply of power must increase to maintain the voltage. When this fails, the whole grid collapses. In our technological society, that creates a lot of problems, but the collapse of the voltage is standard in the brain although it is quickly restored. When there are more short circuits than power stabilisers, the voltage of the next neuron collapses. If there are more of the stabilisers (inhibitory synapses), the voltage is maintained, and nothing happens. When the voltage collapses, a new nerve impulse is initiated at the beginning of the axon running along at 100–200 km/h to the next synapse where the whole game is repeated. At each synapse, excitatory and inhibitory signals are weighed up against each other to make another decision. It is not difficult to imagine that this makes a powerful computer. John Carew Eccles (1903–1997), who was initially opposed to the idea of chemical synapses in the brain, convinced himself of the opposite by carefully studying inhibitory synapses and then became a fervent proponent of the chemical synapse. This finally ended the war of soups and sparks and eventually earned him the Nobel Prize for his insights about the processing of excitatory and inhibitory nerve inputs in 1963.

Now the reader may say that nerve impulses jumping from neuron to neuron is like a chicken-and-egg scenario because you always need one neuron to excite another one. Where and how does it start?

This is where our senses come in. Our eyes see something which is translated into a short circuit that initiates a cascade of nerve impulses from our eye to the part of our brain that deals with visual information. From there on it could trigger a decision to wave a hand because we saw a person we know. The 2021 Nobel Prize in Physiology and Medicine was awarded to David Julius and Ardem Patapoutian for their studies on how temperature, touch, and pain are sensed and how they trigger the initial nerve impulse. The 2004 Nobel Prize was given to Linda Buck and Richard Axel for the discovery of how smell is perceived. Earlier in 1967, the Nobel Prize was awarded for the primary physiological and chemical visual processes in the eye to Ragnar Granit, Haldan Keffer Hartline, and George Wald. In addition, we have autonomous functions, such as the day and night rhythm, which can generate nerve impulses to regulate our sleep and organ function. The 2017 Nobel Prize was awarded to Jeffrey C. Hall, Michael Rosbash, and Michael W. Young for their work on mechanisms that control our circadian rhythm.

We have seen that the brain uses deliberate short circuits to transmit information. These take only a millisecond but occur in millions or billions of neurons at any given time. After each short circuit, the voltage must be restored; and for this process, our brain needs so much ATP. During each short circuit, ions (sheep and cows) cross the cell membrane, discharging the battery in small steps. Our brain is not in the habit of letting the battery run low; it recharges immediately. In our analogy, we had some sheep (sodium ions) left inside the cow paddock and some cows (potassium ions) left in the sheep paddock after briefly opening the two gates. To restore the system, we can bring them back through a revolving door, which only lets cows move from the sheep to the cow paddock and sheep in the opposite direction (Figure 35). A revolving door does not really exist in our cells; it is rather a sophisticated kissing gate that we met earlier, but the end result is akin to a revolving door, and I will use the analogy here because it is more intuitive. To turn the revolving door, ATP is required. In our neurons another nanomachine, called the sodium pump, moves ions back and forth. It is a close relative of the calcium pump, which we met as the key factor to relax muscle in chapter 3. The sodium pump returns any sodium ions (sheep) that rushed into the cell during a short circuit to the outside and returns any potassium ions (cows) to the cell. Inside our cells we have very few sodium ions but a lot of potassium ions (cows) instead. It is the flow of these two ions that regulates the voltage in a neuron (Figure 35).

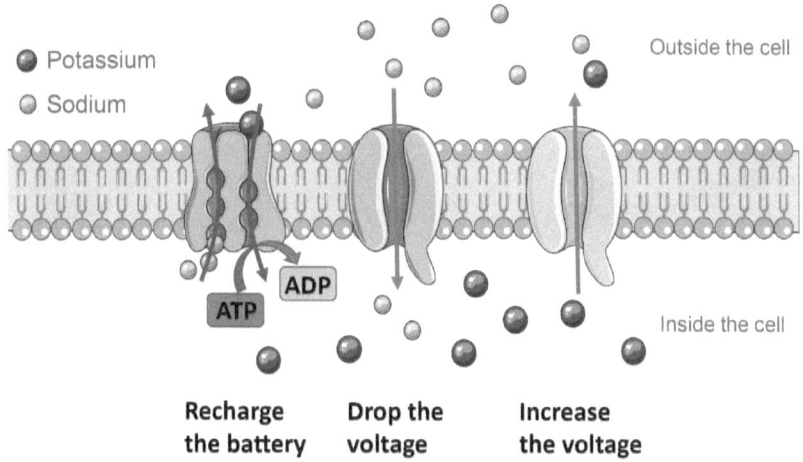

Figure 35. Ions control the voltage of brain cells. An influx of sodium ions (Na) drops the voltage (short circuit), an outflow of potassium ions (K) increases the voltage, and the sodium pump recharges the battery by restoring the original distribution (Servier Medical Art).

The brain uses so much energy because we have sodium ions rushing into millions of cells to drop the voltage and potassium ions rushing out of millions of cells to restore the voltage. The balance between the two decides the voltage of the cell. To keep the battery charged, the sodium pump does the opposite, returning sodium ions back into the brain liquid and potassium ions back into the cell. Each time three sodium ions and two potassium ions are returned an ATP is split to ADP, which again must be restored quickly to ATP by burning glucose. As a result, our brain can only stay conscious for fifteen seconds without blood supply.

How was the sodium pump discovered? Alan Hodgkin and Roger Keynes showed in 1955 that ions were flowing against their concentration gradient between the squid giant axon and its surrounding fluid.[85] At around the same time, Jens Skou (Figure

36) investigated an ATP-using activity in nerve cells from crabs. To split ATP, sodium ions and potassium ions had to be present. This was the result of a serendipitous discovery when he used an ATP preparation containing sodium ions and an ATP preparation containing potassium ions.

Figure 36. Discoverers of the ATP-driven sodium pump: Jens Christian Skou (left, Wikimedia Commons) and Robert L. Post (right, Vanderbilt University Archives).

In our analogy, the revolving door would only turn when at the same time a sheep and a cow pass through in opposite directions. However, Jens Skou initially thought the ions only had to be present for the activity of the sodium pump and did not realise the movement until he read the publication by Hodgkin and Keynes. Moreover, Jens Skou met Robert Post at a conference in Woods Hole and drove with him to Montreal. Robert L. Post (1920–2021, Figure 36) had studied the movement of sodium and potassium ions in red blood cells and found that they move in opposite directions (through the revolving door). He also knew that you could stop this process by using the plant poison ouabain. We met this foxglove poison in chapter 4, where it increased the strength of the heartbeat at lower doses through slowing down the sodium pump.

In 1960, Hodgkin and Keynes showed that when cellular ATP production was poisoned, the sodium pump stopped, but its function could be restored by injecting ATP. This showed convincingly that nerve cells need ATP to function.[86]

Back to neurotransmitters. There are three types of neurotransmitters in the central nervous system. The first type mediates the fast connections to let the nerve impulse jump from neuron to neuron (excitatory). The second type inhibits the propagation of nerve impulses (inhibitory). The third type modulates the sensitivity of many neurons, dampening or enhancing excitation and working much slower than the first two. The common amino acid glutamate, which we use as a flavour enhancer, is the major neurotransmitter of the first type. This is one reason why we need a blood-brain barrier, because our whole brain would die at the glutamate concentration found in our blood. As mentioned earlier, the blood capillaries in the brain are very tight, not allowing blood chemicals to seep into the brain. Some people nevertheless develop headaches after visiting Chinese restaurants because glutamate is used as a flavour enhancer in many dishes and can seep into the brain at excessive concentrations. Neurotransmitters of the second type are aminobutyric acid and the common amino acid glycine. Representatives of the third type are serotonin, dopamine, and noradrenaline which regulate arousal, attention, mood, appetite, etc. Because glutamate is used as a neurotransmitter by the brain, its concentration in brain fluid is extremely low. Removal is achieved by kissing gates, also known as transporters. These transporters indirectly use the energy provided by the sodium pump. Thus, ATP is not only used to keep the electricity running but also to return neurotransmitters to the cells to stop the signal when they have been released. We will discuss the link between the sodium pump and transporters in more detail in chapter 7, but it is the same principle as the removal of calcium ions in muscle to end muscle contraction. Glutamate is used by 90% of all neurons in the brain, and the energy costs associated with re-establishing voltage and pumping back glutamate is thought to account for 80%

of the energy budget of the brain.[87] In the brain, ATP is generated by the oxidation of glucose; and as a result, a steady supply of blood must be ensured to provide oxygen and glucose. There are dedicated gates for glucose in our brain blood capillaries to allow a steady flow of glucose into the brain without leaking anything else.

It took a long time to establish glutamate as the major neurotransmitter in the central nervous system.[88] The molecule was known to have a vital role in metabolism and was initially not considered a suitable candidate. Takashi Hayashi found in 1954 that injection of glutamate into brain produced convulsions. Akira and Noriko Takeuchi were the first to identify glutamate as a neurotransmitter in crayfish muscle in 1964. David Curtis (born 1927) at the Australian National University then showed strong electric action of glutamate in the spinal cord but was not yet convinced that it was a neurotransmitter. Acceptance grew over many years by increasing evidence of the potency of glutamate and particularly some of its chemical analogues.

Remarkably ATP is not only used to energise the cell, but also used as a neurotransmitter.[89] It belongs to the third group that modulates the activity of other neurons. The regular and rhythmic movement of the intestine is regulated by nerve cells acting on smooth muscle. Smooth muscle is different from skeletal muscle and serves crucial functions like regulating blood pressure and peristaltic movement of the intestine. Geoffrey Burnstock (1929–2020) proposed in 1972 that ATP was used as a neurotransmitter to regulate smooth muscle function, not only ATP itself but also adenosine. ATP as a neurotransmitter regulates breathing, heart rhythm, and gastrointestinal action.[89] Accordingly, specific receptors have been identified that bind ATP. Mice lacking a specific type of ATP receptors are incapable of tasting, demonstrating its role as a neurotransmitter. It is not too surprising that ATP is also involved in pain perception. Pain is often associated with cell damage, resulting in ATP spilling out of the cell and acting on other neurons. The brain even uses ATP release under physiological circumstances to regulate the speed and sensitivity of nerve fibres.

Identifying the receptor on the other side of the synapse turned out to be even more difficult. The first to develop this concept was John Newport Langley (1852–1925) who also foresaw chemical transmission. He wrote, 'In all cells two constituents at least must be distinguished, (1) substances concerned with carrying out the chief function of the cells [neurotransmitters], such as contraction, secretion, the formation of special metabolic products, and (2) receptive substances [receptors] especially liable to change and capable of setting the chief substance in action.' He found that in muscle he could produce contraction when he applied nicotine directly to the region near the nerve endings where we now know the receptors are located,[84] but not when applied further away. Nicotine is a chemical compound that mimics acetylcholine. Foreseeing chemical transmission, which caused so much controversy much later, he wrote in 1906, 'This seems to require that the nervous impulse should not pass from nerve to muscle by an electric discharge but by the secretion of a special substance at the end of the nerve, a theory suggested in the first instance by du Bois Reymond.' We met Emil du Bois-Reymond as the person who first measured the voltage of neurons. Although the notion of drugs binding to the surface of cells was acknowledged for many years, it took until 1970 to identify the acetylcholine receptor with the help of two tools. The first was the use of the electric organ of the electric ray *Torpedo marmorata*, which is made up of stacks of the receptor. The second was the use of a cobra toxin, which immobilises and suffocates its victims by binding to the receptor on the surface of muscle cells.[90] The electric organ of the ray developed from muscle cells by converting them into a stack of membranes with embedded receptors. When acetylcholine binds to these receptors, they all open and let ions pass through the membrane, which will change the voltage. Because they are all stacked up, the tiny voltages add up to an impressive 70–80 V, which can stun victims close to the ray. The receptors for adrenaline and other hormones and neurotransmitters were identified later. Alfred Gilman (1941–2015) and Martin Rodbell (1925–1998) received the Nobel Prize in 1994 for this achievement. Finally, the genes for these receptors were identified and eventually their structure resolved. Brian Kobilka and Robert J. Lefkowitz received the Nobel Prize in 2012 for these next milestones in receptor research.

We should also briefly talk about the release of neurotransmitters, which also involves ATP. Neurotransmitters come prepacked in small soap bubbles that are located close to the outer membrane of the cell (Figure 34). The bubbles are ready to fuse with the bigger soap bubble that makes up the outer membrane of a cell. It is a bit like a small droplet of water readily joining a larger puddle the moment they touch. A bit different as well in that there is a barrier that needs to be overcome for the final fusion. When the neuron is resting, the bubbles just sit there and wait. When a nerve impulse comes in, the voltage drops, and that opens some floodgates. Like in muscle, calcium rushes into the cell. This is the final trigger to fuse the soap bubbles with the outer membrane, and the content spills out into the cleft between the two neurons. This series of events was first proposed by Bernard Katz (1911–2003) and Ricardo Miledi (1927–2017) in 1965.[84] Like in muscle, ATP is required to mop up the calcium, but there is more. Moreover, ATP is also required to release the soap bubbles which are tethered to ropes to prevent accidental release. Paul Greengard, whom we will meet later, found how ATP flicks a switch to release the soap bubbles.

ATP is also required to package the neurotransmitters into the soap bubble in the first place. Principles and ideas are often reused in biology. The soap bubbles act like a battery, and charging the battery requires ATP. In chapter 3, we saw how ATP is made inside mitochondria. The radial engine, driven by the battery force, pushed ADP and phosphate together in a cylinder to make ATP. In the microworld of molecules, this process can also work in reverse. Thus, splitting ATP revolves the radial engine, which then pumps protons across the soap bubble charging it. In 1981, Yoshinori Ohsumi (born 1945) and Yasuhiro Anraku discovered that reversing the molecular radial engine charges the small cellular soap bubbles, allowing them to take up neurotransmitters using the battery force generated by ATP. Yoshinori Ohsumi went on to win the Nobel Prize in 2016 for his work on how specific soap bubbles in our cells are used to recycle cell materials.

Once pre-packed, we now have hundreds of neurotransmitter packages waiting for delivery. Each time a nerve impulse comes in, a bunch of the packages are released (Figure 34). However, there are sometimes small voltage drops in a neuron, and only one or two packages are released. Bernard Katz (1911–2003) worked out that neurotransmitters are released in packages and received the Nobel Prize in 1970 for this insight. Bernard Katz is another emigre who was raised in a Jewish family in Germany before the war. He studied at the University in Leipzig where he was already awarded a prize for his work on nerve cell function but was not allowed to accept it because of his Jewish background. In 1934, he met Chaim Weizmann, the future president of Israel, who helped Bernard Katz to move to Britain in 1935 where he started to work with Archibald Hill of muscle fame.

We have now established the electrical nature of the brain. A sensory stimulus, such as a sound, vision, or mechanical impression, generates the initial voltage drop, which is propagated to the brain. There the electrical signals are analysed, and a response is generated that involves inputs from many more neurons, generating a perception and a response. Each time neurotransmitters are released from one neuron, they bind to a receptor on the next connected neuron. After integration of all inhibitory and excitatory signals, a new nerve impulse is generated or not. In the brain, specific areas are responsible for tasks as vision, hearing, reason, fear, pleasure, movement, etc. As a result, different areas work hard and use a lot of energy at various times. This has important consequences for the blood flow in the brain and resulted in the development of technologies that can monitor brain activities, which we discuss next.

As I outlined in chapters 1 and 2, the second half of the nineteenth century saw the discovery of the conservation of energy as an important thermodynamical principle.[83] Even psychiatrists were influenced by energy principles at the time. Theodor Meynert (1833–1892) suggested that when energy was used in one part of the brain, an equal amount of energy must disappear from another part of the brain. He further proposed that this was implemented by increasing the blood flow to active areas of the brain and reducing it to others. At the same time, Angelo Mosso (1846–1910) was the first to measure pulsations in the cortex of patients with skull defects. He inferred that during mental activity, blood flow increases to active areas. Hans Berger, whom we met as the inventor of the EEG, used Mosso's technique to confirm these observations. In 1890, Roy and Sherrington observed that local brain volume increased when nerves were electrically stimulated. They proposed 'the existence of an automatic mechanism by which the blood-supply of any part of the cerebral tissue is varied in accordance with the activity of the chemical changes which underlie the functional action of that part'.[91] These studies founded the new field of neuroimaging, but for many years, imaging of the brain could only be performed under certain circumstances and was quite invasive.

This only changed in the twentieth century. Functional magnetic resonance imaging (fMRI), which is based on MRI (which I will explain in a moment), has generated a revolution in our understanding of brain function. It is based on the loading of haemoglobin with oxygen, which can be detected outside the brain when a very strong magnetic field is applied.[92] As we saw in chapter 4, haemoglobin brings oxygen to the tissues where oxygen unloads and diffuses into tissues including the brain. Oxygen-free haemoglobin disrupts the magnetic field and can therefore be detected as a reduction of the signal. This was initially found by Linus Pauling and Charles

Coryell who observed in 1936 that oxygen-rich blood had different magnetic properties than oxygen-depleted blood.[91] The disruption of the magnetic field can be observed when the blood flow to active brain areas is increased to provide additional glucose. The increase in blood flow and glucose consumption occurs earlier than the increase in oxygen consumption. Thus, more oxygenated blood is reaching an active brain area before oxygen is unloaded. This short delay between glucose and oxygen consumption is possible due to the lactate credit card system like in muscle. The difference between muscle and the brain is that lactate is used up locally inside the brain instead of relying on the liver to convert it to glucose. In a further similarity, brain astrocytes contain glycogen, and this glycogen can be used as a readily available store of glucose. As we found out, active areas of the brain need a lot of ATP to reverse ion movement and to return neurotransmitters, which increases glucose consumption. This in turn increases the blood flow to this area to deliver glucose and oxygen. When we think, only certain areas of the brain are active. There are specific areas for visual experiences, hearing, smell, and other sensory inputs. Also, the processing of information does not occur in only one place. Fear, arousal, and critical analysis occur in distinct parts of the brain. As a result, we can now place people in a large magnet and see where oxygen is used to make ATP.

The development of MRI machines, which detect water and on which fMRI machines are based has been surrounded by a lot of controversy, and I am following the account outlined by science writer Morton Meyers.[93] MRI is based on another technology called NMR (nuclear magnetic resonance). Nuclear physicists recognised that protons and neutrons, which can be found in the nucleus of atoms, are spinning and thereby form magnetic fields. Vice versa, when a strong magnetic field is applied, the neutrons and protons will line up like compass needles in the direction of the field. At this point, radio waves can be used to let the spinning nuclei tumble a bit while spinning. When the radio waves are turned off, the tumbling protons and neutrons generate radio waves themselves which can be detected. Chemists have been using this method for many years to determine the structure of organic molecules, because hydrogen atoms can be readily detected. Water has two hydrogens, and these can be detected by NMR. Raymond Damadian (1936-2022, Figure 37) was the first to recognise the potential of this method to detect water in different tissues. Cancer tissue is often very tough while functional tissue is much softer containing more water. Damadian reasoned that this difference could potentially be detected by NMR in an intact organism. A regular NMR machine only holds a pencil-size glass tube containing a chemical compound, and such a machine is already large needing a dedicated room for it because of the large electromagnet required for its operation. Damadian was a physician and a biophysicist and knew the clinical and physical side of the problem. He proposed and showed for the first time that water in cancer tissue behaved differently from water in soft tissues or water in bones.

In 1971, Damadian published his findings in the prestigious journal *Science* entitled 'Tumor Detection by Nuclear Magnetic Resonance'. Paul Lauterbur (1929-2007, Figure 37), an NMR expert, saw the publication and had the next breakthrough insight that the magnetic field could be arranged in a certain way to generate images. Damadian, by contrast, had relied on moving the organism millimetre by millimetre across the magnet and thereby generating an image-like map. Lauterbur also teamed up with mathematicians to generate the software to convert the tiny radio signals into images. Immediately an intense rivalry between Damadian and Lauterbur developed. Damadian was particularly concerned whether he would receive enough credit for his initial observation and started to build his own prototype of an MRI machine. Lauterbur, however, made the next step and published the image of two glass tubes of water. Despite the sophisticated mathematical tools, his team used a pencil eraser to smooth out the first fuzzy picture. Lauterbur sent his publication to the *Science* rival journal *Nature*, where it was rejected as being 'rather trivial' and 'not of sufficiently wide significance'. Lauterbur revised the manuscript and finally got it published in *Nature* in 1973. Thirty years later, the journal counted this landmark publication as one of the twenty-one most influential scientific papers of the twentieth century. Peter Mansfield (Figure 37, 1933-2017) at the University of Nottingham had developed the same technique independently and published an NMR image of a human finger in 1977, while Damadian published an image of a mouse with a tumour in 1976. Damadian then set out to build the first MRI machine that would fit a human. This was a monumental task because the electromagnets required for imaging are massive and need to be cooled by liquid helium to achieve a superconducting state. The researchers spooled 50,000 m of wire to generate the required magnetic power.

Figure 37. Pioneers of medical imaging. (Nationalmedals.org), Peter Mansfield (top), Paul Lauterbur (bottom left), Raymond Damadian, and Lawrence Minkoff (bottom right).

In 1977, Damadian's group finally got a fuzzy cross-section image of a human chest. Damadian wanted to image his own chest, but he turned out to be too large and was replaced by Lawrence Minkoff (Figure 37) from his group. In the same year, Mansfield produced a major leap forward by being able to analyse images from a breathing object and by acquiring images 10,000 times faster. At this point, medical imaging companies became interested in the technology and recruited some of the key scientists to build the first commercial machines. Damadian started his own company which produces MRI machines to this day. However, Paul Lauterbur and Peter Mansfield received the Nobel Prize for their progress towards MRI in 2003. MRI is now a billion-dollar industry, and 60 million scans are performed worldwide each year. Damadian felt that he was overlooked and launched a long-standing media campaign to right the wrongdoing without success. Damadian had a difficult personality and behaved increasingly combative in the competition with Paul Lauterbur. As a result, he was sidelined by the NMR community.

In the remaining chapter, I want to investigate two more questions. First, is ATP required to make memories? And second, do we have more ATP when we feel energised and aroused?

There are distinct types of memories, and it would lead too far away from the topic of this book to go into the details where memories are located and what type they are. However, neurons in an area called the hippocampus are particularly active when certain types of memories are laid down and called upon. When we learn something, we initially store the information as a group of neurons talking to each other. As we saw earlier, a single neuron decides about firing, based on the integration of many different inputs from surrounding neurons, which is the same as information. To store such a pattern, we memorise and recall the information repetitively. This causes a particular pattern of neuronal communication to be reinforced. The brain cannot grow new neurons, but it can improve and strengthen existing connections. It can release more neurotransmitter and grow more synapses where they are required and trim where they are not. A key molecule involved in this process is cyclic AMP.[94] Cyclic AMP is generated when ATP loses two phosphates and then makes a chemical bond with itself. This message is used in many cells of the body to start certain metabolic programs. It can mobilise fat for exercise, and it can initiate the breakdown of glycogen in muscle to provide energy. Earl W. Sutherland (1915-1974) discovered cyclic AMP and suggested in 1960 that it acts as a secondary signal when adrenaline is released during exercise. He received the Nobel Prize in 1971 for his breakthrough in our understanding of how hormones send signals to cells. We will cover this in more detail in chapter 8. In the brain, cAMP tells neurons to grow more synapses, but only in those neurons that fire frequently. To build another synapse costs a lot of energy, and plenty of ATP is required for that. The cyclic AMP uses only a tiny fraction of the available ATP, but it is a very potent signal.

Figure 38. Pioneers who studied the role of cyclic AMP in the formation of memories and nerve transmission: Eric Kandel (left, Wikimedia Commons) and Paul Greengard (right, Wikimedia Commons).

Thus, ATP has two roles in memory formation: firstly, to generate a signal (cAMP) to make better and more synapses and secondly, to provide the energy for the actual building of a new synapse.

A pioneer in this field is Eric Kandel (Born 1929, Figure 38) who received the Nobel Prize for Physiology and Medicine in 2000 for his research into memory formation. He had the insight to use a simple model system, namely the giant sea snail *Aplysia californica*. For his experiments, he used a reflex of the snail, which withdraws its gills when poked. This reflex can be used for a simple learning paradigm called conditioning. During conditioning, a soft touch – which would not result in the withdrawal of the gills – is followed by a stronger different shock. As a result, the snail learns to withdraw its gills already on the first soft touch. The neurons in the snail are very large, allowing the researchers to study the electrical and biochemical processes of this learning process.

Eric Kandel grew up in a Jewish family in Vienna before the war. After the annexation of Austria in 1938, life for Jewish families quickly became difficult. His father was forced to scrub the streets of Vienna with a toothbrush to remove political graffiti in favour of an independent Austria. Eric Kandel was expelled from school, and his parents' shop was given to a non-Jewish citizen. His parents arranged for visas to the United States. Eric Kandel left Austria in early 1939, and his parents joined later in that year.[94] Kandel performed his research on memory formation in *Aplysia* first at New York University and later at Columbia University.

Now to the second question whether ATP is particularly high when we feel energised and aroused. Here the answer is a simple no. As we discussed, ATP is so essential for our brain that its levels are kept up high all the time. When we feel alert and energised, it is due to the activity of a small group of neurons that release noradrenaline.[95] This is the same molecule as the accelerating stuff Otto Loewi discovered in the heart. Otto Loewi did not know the chemical identity of the accelerating neurotransmitter. This was proven by Ulf von Euler (1905-1983) in 1946. At that time, adrenaline was already known as a hormone, but it had also been established that a similar and far more potent compound existed that raised the heart rate. Once available as a pure chemical, the identity of the accelerating stuff was settled. Ulf von Euler received the Nobel Prize for this and other discoveries in 1970.

Noradrenaline belongs to the third type of neurotransmitter which modulates the activity of other neurons in the brain. There are only 32,000 neurons in the brain that release noradrenaline, and they all sit in one tiny spot. However, their nerve fibres reach every part of the brain in large loops and have many synapses along the way. When these neurons are active, they modulate and optimise the fast transmission of signals through receptionists that sit on the surface of many other neurons. Thus, processing is faster, and incoming information is readily processed. There must be the right amount of noradrenaline though. We all have experienced that when you are called to the front of the class to explain something, the neurons are blocked, and nothing comes to mind. Alertness requires the right amount of noradrenaline, and it must come at the right moment. Constant flooding is associated with poor performance and lack of attention.[95] Too little noradrenaline makes you sleepy because other modulatory neurotransmitters are becoming more prominent. Formula 1 drivers probably have an exceptionally low baseline release of noradrenaline and need to do something risky to get them aroused and alert. Vice versa, someone with a higher base level of noradrenaline may find risky behaviour unpleasant because too much noradrenaline is associated with poor performance. The removal of released noradrenaline is a critical step in quenching arousal and requires ATP albeit indirectly. This mechanism is different from the mechanism observed in muscle. Acetylcholine released from muscle nerves is quickly broken down into two parts. Remember, this is the process that was poisoned in Alexei Navalny. Julius Axelrod (1912–2004) worked out that rapid removal of neurotransmitters from the synapse occurs by active uptake back into the cell through a kissing gate. This is an alternative mechanism of quenching neuronal excitation, and he received the Nobel Prize in 1970 for this discovery.[96] This has important practical consequences as reducing the removal of noradrenaline would prolong a state of arousal. This is one of the actions of antidepressant drugs. Noradrenaline reuptake inhibitors, as they are called, are prescribed for attention deficit hyperactivity disorder and depression. Another neurotransmitter that modulates the activity of our nervous system is dopamine. It is released when we find certain activities pleasurable, such as eating, drinking, and gambling. Drugs of abuse target dopamine receptors and dopamine transporters to extend the feeling of pleasure. Dopamine was recognised as a neurotransmitter by Arvid Carlsson (1923–2018) who was awarded the Nobel Prize in 2000. We saw earlier that cyclic AMP derived from ATP can stimulate growth of extra synapses to reinforce neuronal circuits. But cyclic AMP modulates neuronal activity at many places in the brain. Binding of dopamine to its receptor also generates cyclic AMP, and there it is used to modulate the fast nerve transmissions that generate the pleasurable feeling. Paul Greengard (1925–2019, Figure 38) was awarded the Nobel Prize in 2000 for his insights into the actions of dopamine inside neurons. We met him earlier because he found that ATP flicks a switch to release neurotransmitter packages.

While ATP is not involved in arousal, it plays a key role in sleep. Not ATP itself, but ATP stripped of all its energy that is the three phosphates. The remaining molecule is called adenosine (see Figure 1). In 1954, Feldberg and Sherwood found that injection of adenosine into cats' brains caused 30 min of sleep.[97] ATP is kept within cells, but when AMP is formed during intense energy demand, it can be further converted to adenosine, which can leave the cells through a kissing gate. Vice versa, adenosine can enter a cell and be converted back into ATP. Blocking this process prolongs sleep.[98] A genetic variant in humans that reduces the ability to break down adenosine enhances deep sleep.[99] Sleep deprivation increases brain levels of adenosine while deep sleep is accompanied by removal of adenosine. The packages of neurotransmitters that are released from neurons during action often contain ATP as well, which is then broken down into adenosine in the intercellular space of the brain. It seems peculiar that AMP which is generated during intense brain activity can be converted to adenosine as if our neurons were saying that they have worked enough for the day. Maybe ATP usage is the odometer that tells us we should have a rest. There is a specific receptionist that takes the message that adenosine has arrived, prompting neurons to sleep. Certain drugs can trick the receptionist into believing that a large amount of adenosine is present (in pharmacology they are called adenosine receptor agonists), putting neurons into solid sleep. If the receptionist is missing in a particular strain of genetically modified mice, the neurons stay awake. The specific area that puts us to sleep sits above our eyes. Like the neurons that release noradrenaline, these neurons have nerve fibres expanding to many areas of the brain where they release acetylcholine to produce wakefulness. Adenosine blocks the release, thereby putting us to sleep. Caffeine, on the other hand, distracts the receptionist for adenosine (in pharmacology they are called adenosine receptor antagonists) so the adenosine message does not get delivered. There are more regulators of our sleep and wake cycle, but they are not related to ATP, and as a result, we will not go into further detail here.

Now that we have investigated the multiple roles of ATP in the healthy brain, we will also have a look at what happens when there is a catastrophic energy crisis.[100] A stroke happens when a blood clot blocks one of the arteries that provide the brain with oxygen and glucose. Most often the middle cerebral artery is affected. Our brain has two hemispheres, and depending on which side the blood clot sits, one-half of the body is affected. For instance, one side of the face droops when the person tries to smile. When the person raises his/her arms, one arm starts drifting downward. Moreover, the person is confused, and the speech is slurry. Like a heart attack, a stroke is an energy crisis this time in the brain. Glucose and oxygen are missing, and as we saw earlier, the brain has little ability to use anything else to generate energy. Astrocytes will be using their glycogen to generate energy, but they cannot transfer that energy to the neurons. Neurons have phosphagen to recharge ATP like muscle,[101] but this will only last a minute or so. We discussed that the maintenance of the battery charge and the removal of neurotransmitters require ATP. We also learned that the brain cannot tolerate exposure to glutamate in substantial amounts, but during a stroke, glutamate levels in the affected areas match or exceed those in blood plasma. When glutamate cannot be removed or even leaks out of cells, it starts to talk to the receptionists on the other side of the synapses (Figure 34). This will open pores, and sodium and calcium ions will be rushing into the cell. This is done sparingly and locally under physiological conditions to fine-tune the nerve impulse transmission, but it overwhelms the system when too much glutamate floods brain areas. The reduced battery charge aggravates the problem and lets even more calcium ions in. Too much of calcium ions in a cell is unwelcome news for the mitochondria. It causes the battery charge of the mitochondria drop, as well. Some oxygen is still around at this time, and the struggling mitochondria start to produce bleach and peroxide-like chemicals because the normal respiration does not work very well. Some neurons only last a couple of minutes with this onslaught before they die. As a result, it is critical to dissolve the blood clot as soon as possible to avoid permanent damage particularly in areas that still receive a small amount of blood flow.

Many Nobel Prizes have been awarded for fundamental discoveries in neurosciences (see Table 4). Not all are directly related to the function of ATP, but due to the extremely high ATP demand for nerve action, it underpins all its functions.

Table 4: Important discoveries of brain functions and energy

Year	Name	Discovery
1906	Camillo Golgi, Santiago Ramon y Cajal	Nobel Prize for the cellular structure of the nervous system
1932	Charles Scott Sherrington, Edgar Douglas Adrian	Nobel Prize for neuronal functions
1936	Henry Hallett Dale, Otto Loewi	Nobel Prize for the discovery of neurotransmitters
1963	John Carew Eccles, Alan Lloyd Hodgkin, Andrew Fielding Huxley	Nobel Prize for excitatory and inhibitory mechanism of nerve transduction
1967	Ragnar Granit, Haldan Keffer Hartline, George Wald	Nobel Prize for physiological and chemical processes in the eye
1970	Bernard Katz, Ulf von Euler, Julius Axelrod	Nobel Prize for storage release and inactivation of neurotransmitters
1971	Earl W. Sutherland	Nobel Prize for the mechanism of hormone action
1991	Erwin Neher, Bert Sakmann	Nobel Prize for analysis of individual ion channels
1994	Alfred Gilman, Martin Rodbell	Nobel Prize for hormone and neurotransmitter receptors
2000	Arvid Carlsson, Paul Greengard, Eric Kandel	Nobel Prize for signal transduction in the nervous system
2003	Paul C. Lauterbur, Peter Mansfield	Nobel Prize for magnetic resonance imaging
2003	Peter Agre, Roderick MacKinnon	Nobel Prize for the structure of ion channels
2004	Richard Axel, Linda B. Buck	Nobel Prize for the discovery of smell receptors

2012	Robert J. Lefkowitz, Brian K. Kobilka	Nobel Prize for the structure of neurotransmitter receptors
2017	Jeffrey C. Hall, Michael Rosbash, Michael W. Young	Nobel Prize for the mechanisms underlying day and night rhythm
2021	David Julius, Ardem Patapoutian	Nobel Prize for the discovery of temperature and touch reception

6

Hibernating Bears

Hunger, prolonged, is temporary madness! The brain is
at work without its required food, and the most fantastic
notions fill the mind. Hitherto I had never known what
hunger really meant. I was likely to understand it now.
—Jules Verne, *Journey to the Center of the Earth*

Adipose tissue is probably the most underappreciated tissue in our body. This is of course because we are eating too much, and we eat very regularly. This would have been different in ancient times where little food would be available during winter and early spring. Humans have survived many periods of starvation even in modern Europe. The most recent event in Europe was the Dutch famine in the winter 1944–45. Due to a German blockade, food ran out quickly in the big cities. The adult rations in Amsterdam dropped below 1,000 calories by the end of November 1944 and to 580 calories by February 1945.[102] The normal calorie intake for an adult by contrast is 2,000–2,500 calories per day. Wikipedia has a long list of famines with many millions of deaths. Probably the most dramatic recent event was the Great Chinese Famine (1959–1961), resulting in an estimated death toll of 15–55 million people. More well-known are the Irish famines of 1740–1741 and 1845–1852, which caused the death of a significant fraction of the Irish population and triggered migration

123

out of Ireland. In 1694, a French official described the situation after two failed harvests as 'an infinite number of poor souls, weak from hunger and wretchedness and dying from want and lack of bread in streets and squares, in the towns and countryside because, having no work or occupation, they lack the money to buy bread'. In short, famine has been a frequent occurrence for humans, and our adipose tissue prepares us for those times.

Economically speaking, your adipose tissue is like your savings account. Whenever you have spare money, you put it into your savings account. If you spend your money as you earn it, you cannot build up a savings account. These days savings accounts do not generate any interest, and the same is true for our adipose tissue. The only difference is that we cannot remove a large chunk of our adipose tissue to buy something expensive in one go. Our adipose tissue only gives money as needed for essential expenditure.

This raises the question of how big the savings account is and how long it will last. In case of famine, that is not easy to say because people will have something to eat, which extends the reserves. However, we know how much ATP we continuously use in our organs, which never stop working. A hunger strike is the deliberate exclusion of any food with the provision of water and minerals to avoid rapid dehydration. How long could it go? Our liver has energy storage in the form of glycogen. While we are sleeping, this provides us with energy, but it runs out after about 12 h. Our muscles have some glycogen as well, but this runs out even sooner as we have seen in the case of the marathon runner, and it is only for the muscles to consume. That leaves protein and fat. An average weight person has about 15 kg of fat and 6 kg of protein, the latter mainly in the form of muscle mass. Fat provides 9 kcal per g (or 9 food calories as they are tabulated), and protein provides 4 kcal per g. That equals to 135,000 kcal of fat and 24,000 kcal of protein. A 70 kg adult uses about 2,000 kcal per day in which case the reserves should last for approximately 80 days. Not all of the protein reserves can be used, because muscle mass is not fully dispensable. That leaves about 2 months for a hunger strike before it becomes life threatening. The longest known duration of fasting was performed by a man with extreme obesity who fasted for 382 days, consuming only non-caloric fluids, vitamins, and minerals, resulting in a 60% loss of body weight apparently without adverse effects.[103]

If you wonder how we know how long we can starve, Ancel Keys and the Minnesota starvation experiment is the study to look up. Out of concerns over soldiers experiencing starvation and civilians on food rations during World War II, Ancel Keys (1904–2004) proposed to put volunteers onto a restricted diet in 1945. In the twelve-week control phase, all volunteers received a generous diet of 3,200 calories a day followed by a 24-week diet on just 1,570 calories a day. During both periods, volunteers had to work and do exercise. In the first twelve weeks of starvation, the men showed a 21% reduction in strength, 18% loss of weight, and 55% decrease in overall fitness.[104] The volunteers became so weak that they could not turn a rotating door in a department store or open the large door to the library but still had to complete 35 km of walking per week.[105] One person had to be removed from the study because he became aggressive and had dreams of cannibalism. At the end of the 24-week period, the volunteers had lost 24% of their body weight. Even essential organs like liver and kidneys shrink in times of starvation.[106] Health deteriorated as indicated by fluid accumulating in tissues, anaemia, low heart rate, and weakness. These are typical signs of starvation because the body cannot produce enough protein. Importantly, the liver cannot produce enough albumin, the most abundant protein in blood. Its presence causes water to move from tissues into the bloodstream. As we learned earlier, water is constantly produced inside the mitochondria and needs to be exhaled. Otherwise, it will accumulate in tissues. Readers will have seen pictures of starving babies with swollen bellies due to accumulating water. Anaemia is the result of reduced capacity to produce red blood cells. The low heart rate is another desperate measure to reduce the energy consumption during starvation. In addition to the physical effects, study members were depressed, irritable over minor issues, and obsessed about food.

The metabolic changes during starvation were further studied by George F. Cahill (1927–2012) later in the twentieth century. He was the founding father of the science of fasting and starvation and investigated, with the help of volunteers, what fuel is used to what extent during different phases of fasting. One of the major problems of starvation is our hungry brain which as we learned runs almost entirely on glucose when we are well-fed. Remember that our heart likes to use fat (or more precisely fatty acids) probably to spare glucose for the brain. Even when we are starving, our body keeps blood glucose up at > 3 mM. In 24 h of fasting, we need to generate 180 g of glucose of which the brain uses 144 g.[107]

Protein is used to make about 40% of the glucose once we run out of glycogen. Only a small part of our fat, called glycerol, can be converted to glucose but not the fatty acids, which make up most of the fat.[n] There is also some lactic acid, which is produced by our red blood cells and muscles. Despite these additional sources, our muscles would disappear very quickly to generate 180 g glucose per day.

So we need a trick to keep our brain ticking and to preserve our muscles while we are starving. The trick is called ketogenesis (Figure 39). This used to be a term that only biochemists would know, but with the popularity of ketogenic diets, ketone bodies are all the rage. What are ketone bodies? When we use fat as a fuel, we break it down into smaller fragments. Using the peloton analogy again, a typical fat molecule has about fifty-five cyclists. In the first step, we separate the group into four smaller pelotons. One of only three cyclists, called glycerol, and three with sixteen to eighteen cyclists, which represent the fatty acids. The three-cyclist peloton leaves the adipose tissue and

---

[n]   To be precise, a small fraction of fatty acids can also be converted to glucose, but this would lead too far away from the topic.

moves to the liver where they join other pelotons of three cyclists to make groups of six, which are then readily converted into glucose.

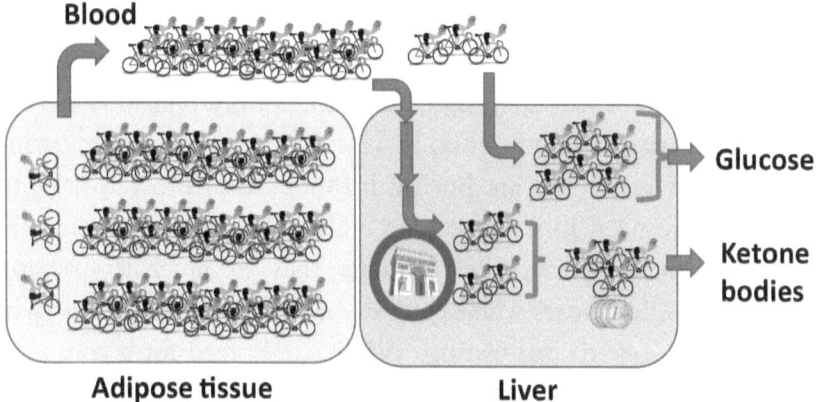

Blood

Glucose

Ketone bodies

Adipose tissue   **Liver**

Figure 39. Ketone bodies as a spare fuel during fasting and starvation. Adipose tissue breaks down fat into fatty acids (large peloton) and glycerol (three cyclists), which are released. Both transfer to the liver where they are used to make ketone bodies and glucose, respectively.

The three larger pelotons also cycle from adipose tissue to the liver, but they are taken apart further into groups of two cyclists. The two cyclists could enter the Arc de Triomphe cycle race, but during fasting or on a ketogenic diet, an alternative is favoured. They can become ketone bodies. For that we group them together to become groups of four cyclists, and we give them some coins as well. Fixed up like that, they leave the liver and go to any other tissue that needs energy, particularly the brain. Normally our brain only uses glucose to generate energy, but during starvation it starts using ketone bodies. Otherwise, we would have to give up too much muscle mass.[108] All this is optimised to allow survival for the longest time possible. After a week of fasting, instead of 180 g, only 80 g of glucose is made per day, and only half of that goes to the brain. Muscle protein contribution to glucose generation is reduced to 25%. Instead, ketone body production provides the remaining energy for the brain and other tissues. A further advantage of switching from protein to ketone bodies is the reduced amount of urea that is produced by the liver. Urea needs to be eliminated through the kidneys, and for that we need water. This helps if the famine also limits access to water.

There is another reason that we are so good at sparing glucose: we are even born on a ketogenic diet. At birth, our brain uses 60%–70% of the total energy bill of our body. Half of that comes from ketone bodies. As a result, we are born a little obese to provide us with all the fat we need to make ketone bodies. Because the brains are comparatively smaller in other animals, they are much less relying on ketone bodies. Bears who hibernate have less ketone bodies than humans after one day of fasting. Whales only feed for a couple of weeks per year and have less ketone bodies than we do.[108]

About 150 g of adipose tissue is used per day during fasting. This is valuable to remember the next time the reader steps on a balance to check weight loss during fasting. The biggest weight loss after the first day of fasting comes from the weight of water stored together with glycogen in liver. There are about 80 g of glycogen stored with 160 g of water. When we lose 300 g on day one of fasting, we have mostly lost glycogen and water but only a small amount of fat. This makes it so difficult to lose fat when we fill up our glycogen with carbohydrates each day. It also explains why we tend to eat too much and gain weight. We rarely had the opportunity to accumulate much fat in prehistoric times, and some fat was certainly a good thing to prepare us for the next famine. While our body is very good at telling us how much we need to consume to keep our energy balance, it errs slightly on the generous side to fill up the savings account.

While adipose tissue has long been recognised as a storage organ, more recently it has also been recognised as an organ that produces hormones. A major hormone that tells our body that we have enough fat storage is called leptin. Mice that lack leptin are monstrously obese, and daily injections of leptin rapidly reduce food intake and body mass. There is a whole host of hormones produced by adipose tissue; but for the sake of our topic, ATP, we will stick to leptin. Leptin production is proportional to the mass of adipose tissue and peaks during the night. Some of the receptionists for leptin are found in the brain where appetite is controlled. Leptin signals to the brain that we have sufficient reserves and can reduce food intake and be happy. Leptin receptionists are also found in other tissues, for example in muscle, where they convey the message to use more fatty acids for energy production.

There is one type of adipose tissue that everybody likes. It is called brown adipose tissue and helps animals keep their body temperature up in cold climates. It is also called the hibernating gland because hibernating animals use it to keep warm. Adult humans generate heat through muscle movement. We shiver when it is cold, and as we saw earlier, heat is generated when ATP is split during muscle movement. Newborns do not move much, and consequently they have some brown adipose tissue, but it disappears in adults, and only remnants are found around arteries passing along the ribs to maintain blood temperature.[109]

How do you make heat without moving? When we split ATP and use it to do some work, the energy released becomes part work and part heat. If we split ATP without carrying out any work, that would just generate heat. In brown adipose tissue, using the mitochondrial battery analogy again, we have twenty times more paddle wheels that generate electricity and charge the mitochondrial battery than radial engines that use the electricity.[110] This would normally not make any sense because the battery charge is used to rotate the shaft of the radial engine to make ATP. In brown adipose tissue, the battery is simply short-circuited; and as in any short circuit, the energy is converted into heat. Fuses in electrical appliances rely on thin wires that melt during a short circuit, thereby breaking the circuit. In brown adipose mitochondria, there is no fuse; and when they are activated, heat will be generated (Figure 40).

The short circuit is mediated by an ion channel or gate, a type of protein that we met in detail in chapter 5. The only difference is the type of ion used for the short circuit. In this case, it is protons or acid, because protons are used to charge the mitochondrial battery. In mammals, heat generation peaks in the first week after birth and decreases continuously after that.[110] In adults, extended exposure to cold temperature increases the capacity of the short circuit. That means we have to move to the poles to lose weight. However, there is hope that some normal white adipose cells that reside under the skin may undergo 'browning' and become so-called beige adipose cells.

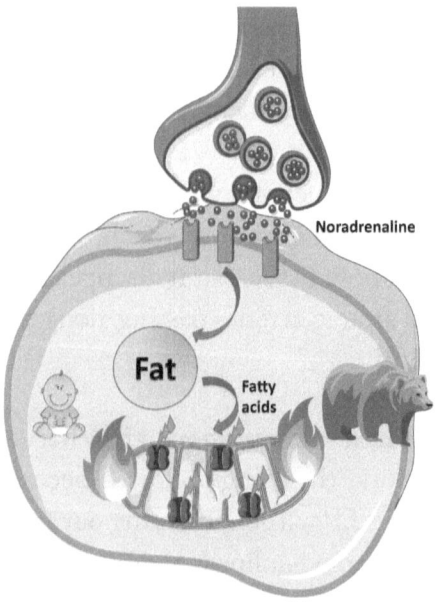

Figure 40. A cartoon of brown adipose tissue. When noradrenaline is released, fat is broken down and the resulting fatty acids used as a fuel to generate heat, which hibernating animals and newborns enjoy. The short-circuiting channels in the membrane of the mitochondria are indicated.

A lot of hype happened when the Harvard biologist Bruce Spiegelman published in 2012 that exercise caused the release of a novel hormone, called irisin, which could turn white adipose cells into 'beige' adipose cells.[111] This was hailed as the 'exercise' hormone that we could all take to lose weight. A start-up company named Ember Therapeutics was set up in due course. The original study was subsequently criticised for technical reasons, and doubts were raised about the real role of the hormone.[112,113] Ember Therapeutics has now changed its focus to other hormones that push the development of adipose cell precursors into brown adipose cells.

In humans, a hibernating gland was described by Shinkishi Hatai (1876–1963) in 1902 in newborns. The role of brown adipose tissue as heat generating was first proposed by Robert Smith in 1961, and the short circuit was proposed by David G. Nicholls in the early 1970s.[110]

Heat generation is modified also by changes of body weight. In an average person, it amounts to about 100 calories per day. In an obese person after weight loss, it can reach up to 500 calories per day. In starving individuals, it becomes negligible.[106]

How does it start? The natural agent to initiate heat production in brown adipose tissue is noradrenaline, the same neurotransmitter that lets our heart go faster and that makes our brain to focus and pay attention. Noradrenaline causes the generation of cyclic AMP, which as we have learned is derived from ATP and which is the signal to start making heat. This can be used to measure how much heat is produced. When noradrenaline is added to a suspension of cells derived from brown adipose tissue, they instantaneously generate 300 W/kg of cells. This is an enormous amount compared to our whole-body heat production of 100 W when we are resting. This of course implies that activation of brown adipose tissue in obese humans could reduce body weight. However, in adults, at normal temperature the amount of energy consumption by brown adipose tissue is too small to make any difference. There is a chemical short circuit though, known as 2,4-dinitrophenol. After ingestion, it constantly discharges the mitochondrial battery in every cell in our bodies. Weight loss is guaranteed, but the therapeutic window is narrow, as they say, which means that it is easily overdosed. This can lead to heart failure because the compound reaches mitochondria everywhere including the heart, thereby increasing the heart rate.[114] The weight-loss activity of 2,4-dinitrophenol was discovered in 1933, and it was soon marketed as an over-the-counter drug for weight loss. Because of its toxicity and narrow therapeutic window, it was labelled as unsuitable for human use in 1938 in the USA and later in other countries. However, many websites still offer it illegally and advertise it as a fat burner. Safer alternatives have been developed that are not prone to overdosing. It remains to be seen whether these will become effective medicines for weight loss.[115]

We have discussed quite extensively how we use fat in times of need, but not how it is made. We put money into our savings account when we have income, typically on payday. The same is true for our body. We only eat every four hours during the day and sleep for eight hours. At least we should eat enough to keep our ATP production up for the next four to eight hours, but often we eat more than that. There are three scenarios that we are all familiar with, particularly above the age of forty, namely the fatty food belly, the spaghetti belly, and the beer belly. The first one is easy to explain. Eating too much fat, think creamy desserts and cakes, will result in storage as fat, no surprise.

How about the spaghetti belly? Justus von Liebig, who discovered the major nutrients carbohydrates, fat, and protein, was the first to notice that nutrients can be converted. He observed that a lean goose weighing 4 pounds gained 5 pounds in weight in 36 days during which it had been fed with 24 pounds of maize. Of this weight gain, 3.5 pounds was in the form of fat, which could not have been derived from the maize as it contained less than 0.1% of fat. Before we convert carbohydrates to fat, we initially store carbohydrates in the form of glycogen. Up to 80 g carbohydrates are stored this way. However, 80 g dry spaghetti, which is mostly carbohydrates, does not seem like a generous portion; but it is sufficient. Secondly, our glycogen is rarely fully depleted. It takes 24 hours of fasting to bring it down. That means that most of our spaghetti portion must be converted to fat stored somewhere else. This happens in our liver where the carbohydrates are converted into fat. From there on they are shipped to the adipose tissue for long-term storage.

The beer belly works slightly different. Alcohol contains a good chunk of calories. We do not notice that as much because drinks do not extend our stomach, which is one of the major signals to stop eating. Going back to our peloton analogy, alcohol is a group of only two cyclists. One of the idiosyncrasies of our metabolism is that we cannot combine three groups of two cyclists to make glucose (six cyclists); we can only convert two groups of three cyclists to glucose. That is also the reason we cannot convert fatty acids into glucose. Instead, our liver converts alcohol into vinegar, and the vinegar can be used by our muscles as fuels. However, we drink in the evening and typically do not exercise after two or three glasses of wine. In that case, the vinegar is converted to fat. In adipose tissue and in the liver, we have the right equipment to combine eight to ten times two cyclists into one molecule of fatty acid. The fatty acids are then combined with a three-cyclist group derived from glucose to make a proper fat molecule. In contrast to carbohydrates, fat can be stored without water. A look into a tab of margarine will confirm this concept. It is insightful to think of weight gain as the number margarine tabs added to your hips and belly. Importantly, there is no limit for storage. Biologically this even makes sense. Before hibernation, bears eat as much as possible to become fat and have enough reserves. Fat bear week is an annual contest in Katmai National Park in Alaska where photos are shot and online voting occurs to nominate the fattest bear (Figure 41). To make fat from scratch, which occurs both in the liver and in adipose tissue, we use a convoluted way of rearranging pelotons. It would be too complicated to cover this in detail, but I can give the kitchen recipe. What we need is a group of two cyclists that is generated from glucose; ATP is required for several steps, and lastly, we need Otto Warburg's coferment.

Figure 41. A fat bear before hibernation.

ATP is required to make certain reactions go forward, which otherwise would not work. The coferment is required to do the opposite of burning, it is adding coins to nutrient fragments. As we discussed earlier, burning is trying to put as many oxygens as possible onto carbon atoms. The opposite is to put as many hydrogen atoms as possible onto carbon atoms. The coferment does exactly that; it transfers two hydrogen atoms onto a carbon atom. As mentioned earlier, a fat molecule is not unlike petrol or candle wax. Both have close to the maximum amount of hydrogen atoms attached to chains of carbon atoms. In our analogy, that is a peloton of sixteen to twenty cyclists. One fat molecule provides energy to make close to 400 molecules of ATP. That makes a nice savings account and can be grown to a generous size.

We understand now how to build up a nice reserve for tough times, but how do we release the fat? Adrenaline is the key. It is released from a gland on top of our kidneys, for example during exercise, and announces itself to the receptionists on the adipose tissue. They convey the message upon which ATP is turned into cyclic AMP. Cyclic AMP initiates a cascade of events which ends in breaking down fat into fatty acids and glycerol. While adrenaline is released during exercise, another hormone called glucagon is released during fasting and causes the same sequence of events. Insulin prevents this from happening. Exercise after a meal will not reduce fat storage, because insulin has been released. To lose fat, exercise should be done before a meal.

I hope the reader will appreciate his or her adipose tissue after reading this chapter. It would take another book to write about the health effects of too much adipose tissue, but this would lead us off the path, and therefore, I will stop here.

7

Mythbusters and Blockbusters

How many things have been denied one day,
only to become realities the next!
—Jules Verne, *From the Earth to the Moon*

The intestine is a very large and often underappreciated organ until it malfunctions. There is the small and large intestine. The small intestine manages the digestion of food together with the stomach, while the large intestine contains most of your microflora and removes the liquid from the remainders of the digested food. For the theme of this book, I will focus on the small intestine. The human small intestine is about 7 m long, and its surface area is a massive 200 m², which is about the same area as a tennis court. The reason it is so large are the folds and fingers that make the surface of the intestine (Figure 42). In addition, each cell facing the lumen of the intestine has finger-like protrusions on its surface. So we have folds, fingers, and mini fingers, the latter being called villi and microvilli, respectively. That allows the tennis court to fit into our belly. Our body has special cells when it contacts the outside world, known as epithelial cells. They are found in the lung, intestine, skin, glands, and kidneys, among other places. Some epithelial cells secrete things, for instance in sweat glands; others absorb things such as in the small intestine. The intestine has also epithelial cells that secrete stuff, such as mucus, to allow the digested material to proceed smoothly. Epithelial cells that absorb or secrete stuff have a ruffled surface on the side that faces the environment and a smooth side where the cells are facing the blood.

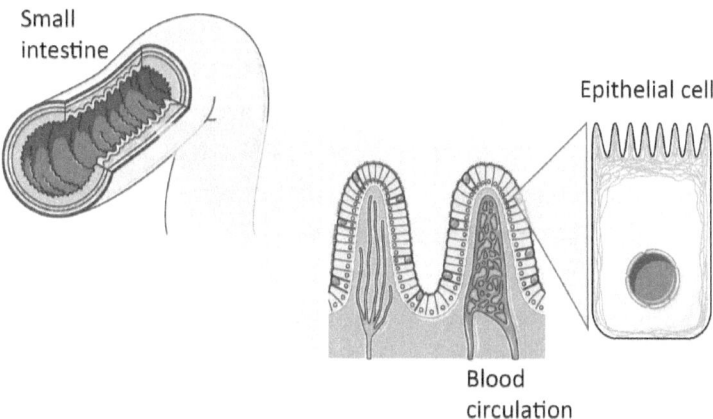

Small intestine

Epithelial cell

Blood circulation

Figure 42. Structure of the small intestine. The intestine has folds and fingers that are lined by epithelial cells. Each epithelial cell has microfingers, which further enlarge the surface of the intestine. Image credit: Servier Medical Art.

As a result, they can easily exchange material between the environment and the blood or vice versa. The enlargement of the surface area of the intestine serves to absorb food efficiently. Our intestine also takes its time, about 24 h, for the food slurry to proceed from start to finish. This makes it quite efficient. Depending on the food source, up to 95% of the available nutrients are absorbed during the passage. Before we can absorb food, we must digest it. Most food items we eat are similar to us, for example, muscle meat from cows, pigs, lambs, and chickens; or they are derived from plants. Plant food is either roots or seeds that store starch, leaves in some shape or form, or fruits. To give shape and form to parts of an organism, a lot of polymers are used. In animal tissues, these are proteins such as collagen and mixed polymers of sugars and proteins. Proteins are made up of twenty odd different amino acids that are strung together like pearls. However, they are not hanging like a necklace but form coils and loops and are tightly packed together (see Figure 23 as an example). Certain proteins are in addition coated with long strings of sugars, which bind water and make the tissue elastic. This is the reason our skin is soft and at the same time tear resistant. Skin is a complex mixture of strings that are interconnected with each other, and therefore, skin becomes very chewy and tough when cooked. Pork crackling shows all these characteristics. To digest such a mesh of proteins and sugars is not an easy task. Plant parts that are starch storage organs are comparatively easy to digest, such as potatoes and rice. Grains are even pre-processed in the form of flour and therefore easy to digest as well. Then there are other parts of plants which we cannot digest at all, particularly the cell walls. These are called fibre, and our microflora deals with those to some extent. Mechanical chewing generates a pulp that is easier to digest. It is not essential because carnivores swallow substantial pieces of meat and let the stomach do the rest. If you have a Labrador at home, you will be surprised what can be digested without chewing. Let us first have a look at what the stomach does and how its function was discovered.

Friedrich Tiedemann (1781–1861) and Leopold Gmelin (1788–1853), respectively professors of physiology and chemistry at Heidelberg, reported in the 1820s that the stomach of a fasting dog contained a neutral liquid, which rapidly became acidic upon ingestion of food.[116] At the same time, William Prout (1785–1850) identified the acid as hydrochloric acid. Tiedemann and Gmelin also discovered that the breakdown of starch in the intestine produced sugar. Because strong acids can break down proteins and starch into its units (or pearls), the process of digestion was considered by many scientists to be a chemical process, not a biological process. The first doubts were raised by William Beaumont (1785–1853), an American military surgeon who had a patient with a gunshot wound that extended into the stomach, allowing samples of gastric juice to be taken. He performed the important control experiment to compare the action on a piece of meat, of acid alone compared to gastric juice. Because gastric juice was much better at digesting meat, he concluded that gastric juice contained some important digestive principle in addition to acid. Nepomuk Eberle (1795–1834) used an acidified extract from dried stomach lining to treat egg protein and concluded that the egg's protein-like properties disappeared. Using Eberle's method, Theodor Schwann (1810–1882) extended these experiments and concluded in 1836 that gastric juice contained a 'ferment' that digested protein, which he called pepsin. 'Ferment', to remind the reader, was an early name for enzymes because they caused a conversion of biological material, like the conversion of sugar into alcohol as carried out by yeast or a yeast extract. These days enzymes that digest protein are called proteases. Many prominent researchers opposed Schwann's idea, but he was supported by Louis Mialhe (1807–1886) in France who also recognised the special action of salivary gland juice on starch and of gastric juice on proteins. The existence of pepsin was firmly established during the 1880s by the Cambridge physiologists John Newport Langley (1852–1925) and John Sydney Edkins (1863–1940). Pepsin cannot be produced as an active digestive enzyme by the stomach because it would self-digest the stomach cells. Instead, it is produced as an inactive precursor, which is activated by acid. The precursor is called pepsinogen (pepsin generator). Langley and Edkins recognised that pepsinogen was stable in alkaline solutions, allowing its isolation. Finally, pepsin was isolated and crystallised in 1930 by John Howard Northrop (1891–1987).[116]

Acid is a powerful ingredient to initiate digestion. The acid denatures protein. Denaturation means that the shape of the protein is lost, but its pearls – amino acids – are still lined up as on a necklace. The denaturation process can be demonstrated by adding citrus juice to milk, which results in curdling. The curdling comes from the fact that denatured protein likes to clump together. Pepsin can now attack the pearl string, cutting it into shorter strings.

Where does the acid come from? In the end from water. When carbon dioxide dissolves in water, bicarbonate and protons (acid) are generated. These protons are pumped out into the stomach. Like the sodium pump and calcium pump, which we already met in earlier chapters, there is a proton pump. It uses ATP to pump protons into the lumen of the stomach.

Sometimes there is too much acid. Heartburn is a symptom produced by reflux when digested food and gastric acid pass back up into the oesophagus. Everybody who has indulged in too much Cabernet Sauvignon may have had that occurring during the night, and it is a very unpleasant sensation. When reflux is becoming a chronic condition, it can result in oesophageal ulcers. As a preventative measure, you can take a pill of Prilosec or a related compound to avoid the build-up of too much acid in your stomach through blocking the proton pump.

George Sachs (1935–2019), working at the University of Alabama, demonstrated the existence of the proton pump in the late 1960s.[117] He initiated a drug discovery project with the pharmaceutical company Smith Kline & French (SK&F) in Philadelphia for the treatment of peptic ulcers. However, in 1973, James Black (1924–2010) and colleagues from SK&F discovered another group of drugs that reduced the production of gastric acid by blocking the receptionist who receives the signal to initiate acid production in the stomach. James Black received the Nobel Prize in 1988 for the development of β-blockers, which are used for a variety of cardiovascular conditions, and for the development anti-acid drugs. This stopped efforts at SK&F to look for other anti-acid drugs. Around the same time, a Swedish company, Hässle AB, which later merged into AstraZeneca, started an independent program to generate new anti-acid drugs. They tested series of compounds in live animals and slowly made progress but without having an idea by what mechanism the compounds might act. In 1977, the team from Hässle met George Sachs at a conference in Uppsala where both sides presented data. After initial hiccups, because of chemical instabilities of the compounds, it became clear that the Hässle drug blocked the proton pump. This allowed to move further development of the new drug away from living animals to biochemical preparations of the proton pump, which is much faster. Toxicity associated with the original drug was reduced in due course, and in 1976, an improved drug called picoprazole was developed. There were initial concerns about the toxicity of picoprazole as well, when tested in dogs. It turned out, however, that the side effects were caused by antiparasitic drugs given to the dogs for unrelated reasons. Finally, after further rounds of improvement, a new blockbuster drug was born in 1979 and given the name omeprazole or known by its trade name Prilosec. Overall, the project had taken ten years from start to finish, and human trials were launched in 1980. In 1988, Swedish authorities approved omeprazole for the treatment of duodenal ulcers and reflux. The drug has an interesting mode of action, because it gets activated by the acid in the stomach and then makes a chemical reaction with the proton pump, thereby blocking it. Most other drugs are just sticking to a pocket in a protein, blocking the entry of a reaction partner. Proton-pump inhibitors in general and Prilosec in particular revolutionised ulcer treatment °. Prilosec became a blockbuster drug for AstraZeneca shortly after its launch. At its peak in 1999, annual sales amounted to $6 billion.[117]

After covering the role of ATP to pump acid into the stomach, we can now return to digestion. As mentioned, pepsin in the acidic environment of the stomach breaks down protein, which we visualised as strings of pearls broken into shorter strings. Initially it was thought that these shorter strings were taken up by the intestine; but Otto Cohnheim (1873–1953), using extracts of intestinal wall, showed in 1901 that the strings are broken down further into individual pearls, which are finally absorbed by the intestine. This suggested that there were more, yet unrecognised, digesting enzymes. Otto Loewi (1873–1961) of 'Vagusstoff' fame discovered in 1902 that nutrition was adequate when protein was replaced by a mix of individual pearls, also known as amino acids. However, due to limitations of analytical

---

° These days we know that most gastric ulcers are caused by a bacterial infection but proton pump inhibitors are still used in combination with antibiotics.

techniques at the time, it was difficult to detect individual amino acids generated by digestion in the lumen of the intestine as well as their subsequent arrival in the bloodstream.

In the search for more digestive enzymes, Tiedemann and Gmelin reported in 1827 that pancreatic juices had digestive properties. The pancreas is a large gland that attaches to the first part of the small intestine, called the duodenum. It has a duct, collecting all the juices it produces and emptying them into the duodenum. The juices contain a lot of digestive enzymes and bicarbonate to neutralise the acid coming from the stomach. Generating a large amount of bicarbonate to neutralise the stomach acid also requires ATP, but this time to drive the sodium pump. Inside the pancreas cells, carbonic acid is generated from carbon dioxide. Bicarbonate is carbonic acid stripped of its acid making proton. Since we only want to secrete bicarbonate to neutralise acid, we must send the proton back. It is pretty much the opposite of what happens in the stomach. The whole operation involves a little dance through kissing gates involving bicarbonate, chloride, sodium ions, and protons. Sodium is imported to get the protons out of the cell into the blood, while bicarbonate is left behind. To restore the sodium balance, sodium ions are returned through the sodium pump energised by ATP.

It took until 1876 before Friedrich Wilhelm Kühne (1837–1900) demonstrated the presence of digestive enzymes in pancreas juice. He initially called the main enzyme pancreatin and later trypsin.[p] In contrast to pepsin, trypsin was active in neutral solutions. This makes sense because we just neutralised the gastric juice with bicarbonate. While trypsin can break down whole proteins, Cohnheim's extracts from the surface of the intestine could only break down shorter pearl strings into individual pearls. This again suggested that the digestion of protein by the stomach was incomplete, and more enzymes were required. He called the associated activity 'erepsin'. Using improved

---

[p] We know today that pancreatic juice contains a multitude of enzymes for the digestion of proteins, fat, starch, and other food components.

analytical methods, Donald Van Slyke (1883–1971) and Gustave M. Meyer (1875–1945) concluded in 1912, 'The increase in amino acids . . . of the blood, noted during absorption of protein, is . . . positive evidence that amino acids as such do normally pass the intestinal wall and enter into the blood current.'[118]

By the 1920s, a picture was emerging that digestion of protein occurs in several steps. In the stomach, the intricate structure of intact protein is destroyed by acid, resulting in curdling. This allows pepsin to attack the strings of pearls that make up protein and release smaller strings. The acidic pulp of the stomach is neutralised by pancreatic juice that also contains trypsin (and other enzymes that attack and split protein), resulting in further shortening of the strings. Finally, Cohnheim's erepsin[q] releases individual pearls, which are finally absorbed by the epithelial cells of the intestine and passed into the bloodstream. The next question to resolve was how individual pearls or amino acids move through the epithelial cells into our blood.

Svante Arrhenius (1859–1927) was the first to recognise in 1909 that sugars and other nutrients did not just pass through the walls of the intestine. It was also observed that some sugars were removed faster from the lumen of the intestine than others and that 'vital' tissue was required. This suggested the presence of an active removal process, and as we have established, one of the properties of 'vital' tissue is the maintenance of ATP. In 1938, Ernst Bárány and Erik Sperber showed that glucose transport across the wall of the intestine could move glucose against a concentration gradient, the definition of an active process. They performed the experiment with a narcotised rabbit. They exposed the intestine and isolated a 50 cm section of it by tying up both ends. When they injected a glucose-containing solution into the section, they found that glucose disappeared within 2 hours, reaching a concentration that was only 1/10 of that in blood. The concentration of another sugar-like nutrient that was not taken

---

[q] Today we know that erepsin is a mix of a handful of so-called brush-border peptidases, which are enzymes that sit at the coal face of the intestine.

up by the intestine remained constant. It took surprisingly long to recognise where ATP was involved in the process of intestinal nutrient absorption. We have seen earlier that the sodium pump returns sodium ions after they rushed into the cell during a nerve impulse. The sudden changes of ion concentrations are a peculiar property of excitable cells in the nervous system.

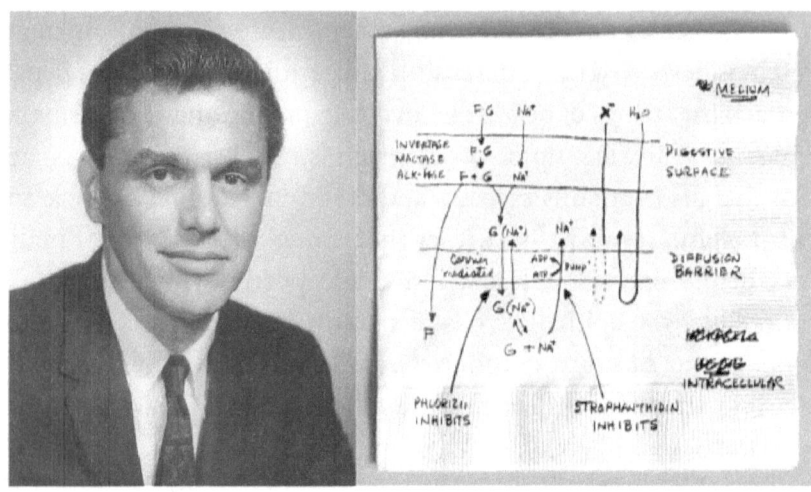

Figure 43. Robert K. Crane and the napkin sketch. The interesting part is the diffusion barrier, which represents the membrane of the cell. Compartments are listed as medium (outside) and intracellular. G symbolises glucose and Na⁺ are the sodium ions. The passage through the kissing gate is symbolised as two opposing arrows, because it can go in both directions. However, when Na⁺ is removed by the sodium pump (just pump in the sketch), glucose is dragged along. Please note that Robert Crane had the sodium pump in the wrong membrane. It should be sitting at the opposite end of the cell, which is not depicted. F is fructose; in his scheme, he starts with sucrose (sugar F-G), which is digested into glucose and fructose (from Ref. [119]).

Most other cells just keep sodium ions low inside the cell and keep the voltage constant. Cell membranes do not let sodium ions in easily, so a large concentration gradient exists between blood and the space inside a cell, and this is maintained by the sodium pump. Robert K. Crane (1919–2010, Figure 43) recognised that this could push the removal of glucose from the lumen of the intestine into the epithelial cells. It is like an escalator in a shopping mall. If we want to go uphill, we step on the escalator, and it drags us along. In this case, glucose hops onto the sodium concentration gradient and gets dragged into the cell where the sodium ion departs. Sodium is pushed out of the intestinal cell on the other end by the sodium pump. Glucose leisurely leaves the cell through another gate without company and enters the bloodstream. Robert K. Crane sketched the idea on a napkin just before giving a presentation at an international meeting on membrane transport and metabolism in Prague in 1960 (Figure 43).[119]

In his view, ATP indirectly drives the absorption of nutrients in the intestine by constantly pushing the sodium escalator, which is used by the nutrients. This is true not only for glucose, but also for amino acids. Peter Mitchell, who developed the idea of the mitochondrial battery, attended the meeting. He immediately recognised the conceptual advance and responded with 'You have got it.'[120] The escalator principle has since been confirmed many times and is observed in all cell types. On top of that, it has a very practical application, called oral rehydration therapy, that has saved millions of lives. In many poor regions of the world, people can die from dehydration during diarrhoea. Water alone will not rehydrate the body, because it is not readily taken up by the intestine. However, if it is combined with salt (sodium ions) and glucose, the water gets dragged along with glucose and sodium ions into the body. Nowadays, we even know that there is a water pore in the glucose escalator.

I want to briefly come back to the glucose escalator. We learned earlier that gates in the cell rather work like kissing gates. While it is intuitive that an escalator can drag us along for an uphill move, it is less intuitive how a kissing gate could drag us along. We stay with the kissing gate analogy and use sheep and sheepdogs as glucose and sodium ions, respectively (Figure 44).

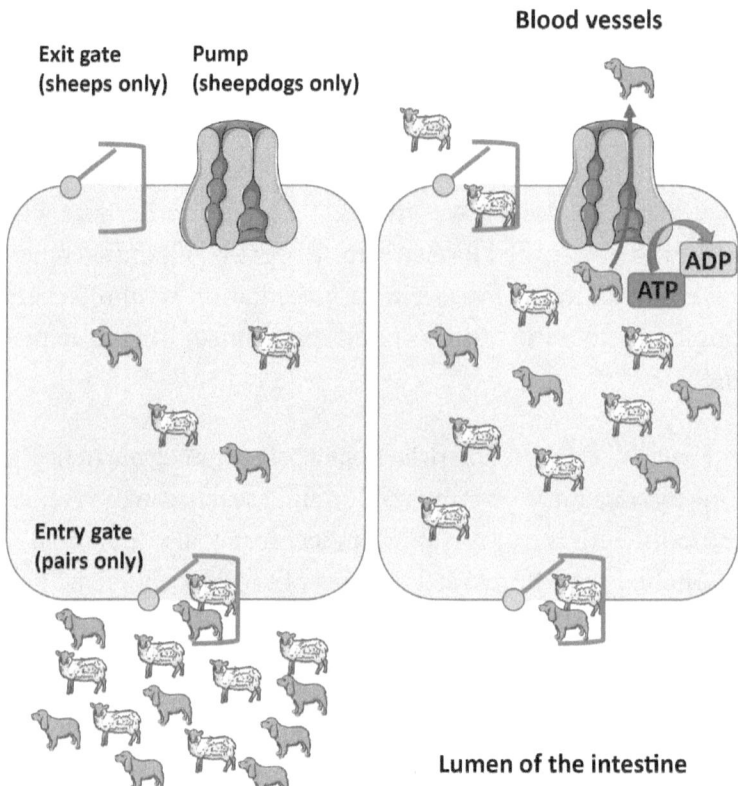

Figure 44. Nutrient absorption. Nutrients are sheep while sodium ions are represented by sheepdogs. For explanation, see text.

The kissing gate is the entry to a paddock. On the outside, sheep and sheepdogs move around randomly, frequently bumping into each other. The paddock has a farmer (the sodium pump) on the opposite end who pushes any sheepdog out that comes into the paddock but he does not do anything with the sheep. On the same side of the paddock, there is another kissing gate, which only allows sheep to leave the paddock. The only rule we need is that the entry kissing gate only let pairs of a sheep and a sheepdog pass. To begin with, we have lots of sheep and sheepdogs on the outside (the glucose that we consumed plus sodium ions from the food and pancreas bicarb secretions) and only very few sheep and sheepdogs inside the paddock. Because of the larger number of sheep and sheepdogs on the outside, it is far more likely that a sheep and a sheepdog bump into each other at the outside entry of the kissing gate. When this happens, they are allowed to proceed into the gate because they are a pair. Once inside the paddock, they drift apart and walk around randomly. Eventually the sheepdog will bump into the farmer who expedites it out of the paddock (the sodium pump). As a result, there are so few sheepdogs inside the paddock that it is very unlikely that a sheepdog and a sheep will ever bump into each other right at the inside of the entry gate. In that very unlikely case, they would go back to the outside of the paddock where they came from. If we continue this game, far more sheepdog-sheep pairs will enter than exit because the sheepdogs are always expedited out by the farmer. Eventually we have enough sheep inside the paddock that they will bump into the second kissing gate at the back of the paddock and leave. At a molecular level, it is just a game of chance that generates the escalator that nutrients take to be absorbed in the intestine.

The key drivers are information and ATP. The farmer only recognises sheepdogs; the entry kissing gate only lets pairs through, the exit kissing gate only sheep. If you find that unlikely, think about a door with a lock and key. Only a specific key can open the lock to the kissing gate. We will come back to information and life in the last chapter. The complete process is driven by the power (ATP) of the farmer who actively pushes out the sheepdogs. Everything else are just chance encounters. The active uptake of sugars into intestinal cells was beautifully demonstrated by William Kinter and T. Hastings Wilson in 1965.[121] They incubated radioactive sugars with inverted sacks of the intestine and captured the radioactivity on film (Figure 45). Because of the inversion of the intestine, the finger-like processes (villi) point outwards and are black because of the radioactive glucose they contain.

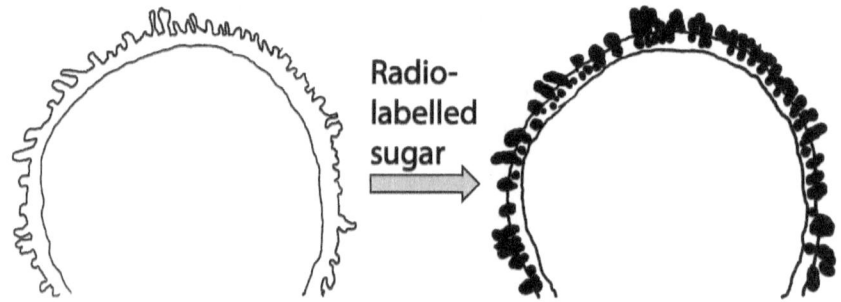

Figure 45. Demonstration of active uptake of sugars into intestinal epithelial cells. An inverted section of intestine was incubated with radiolabelled sugar. The image was captured on film. It becomes black where radioactive material is found. The radiolabel is clearly concentrated inside the fingers that stick out because the intestine was inverted (Drawn after Ref.[121]).

Amino acids derived from protein digestion are absorbed in a similar way as just described for glucose. Specific kissing gates are available for certain amino acids. My own research group identified some of the genes that encode these kissing gates in our genome.

As we saw, nutrient absorption critically relies on the activity of the sodium pump. Sodium is pumped out of cells in all tissues, and the principle of a sodium pump was first postulated in muscle by Robert Dean[122]R.B. in 1941. In 1948, Hans Henriksen Ussing showed that sodium ions were actively moved out of muscle tissue.[123] Karl Zerahn and Ussing then used frog skin as a model epithelial membrane and made use of the electric signal generated by the movement of ions across frog skin. They designed a chamber, which can be used to study ion movements through epithelial barriers, which to this day is known as the Ussing chamber. He also introduced electric terminology into epithelial transport, such as sodium battery and short-circuit current. You probably noticed that I use a similar terminology throughout this book. In 1958, he proposed that in epithelial cells, the sodium pump and entry gates are found on opposite sides of the paddock. When Robert Crane formulated his escalator hypothesis in 1960, he could have placed the sodium pump on the correct side, but perhaps he was not convinced that frog skin was organised in the same way as the intestinal epithelium.

No discussion of the intestine would be complete without looking at intestinal motility. To propagate the food pulp through the digestive system, the intestine constricts in waves that pass the intestine every 80–110 min. Each wave takes 6–10 min to pass a certain point of the intestine.[124] In addition, the small intestine constricts regularly to mix its contents and aid in digestion. This is orchestrated by a nervous system that is found along the intestine which directs the muscle cells that surround the intestine to constriction in the right intervals. The intestinal muscles are different from skeletal muscle in that they constrict much more slowly. They are called smooth muscle cells because they do not show the characteristic stripes observed in skeletal muscle (Figure 21). Smooth muscle cells among other places are found around blood vessels and around the intestine. Their activity is regulated by the intestinal nervous system using numerous neurotransmitters including ATP. Contraction makes use of actin and myosin, the same filaments that are used in skeletal muscle contraction. As outlined in chapter 3, this requires ATP as an energy source, but the contraction is slow and not necessarily triggered by a nerve impulse.

There is another organ where epithelial cells are centre stage; it is the kidney. The kidney is a plumber's heaven. It looks like a large bean from the outside and appears to be smooth, but this is only the outer shell that encapsulates the functional tissue. Inside we find 1–1.5 million tubes (called tubules) that go in loops and end in the centre of the kidney where they merge into the ureter which brings urine to the bladder (Figure 46).

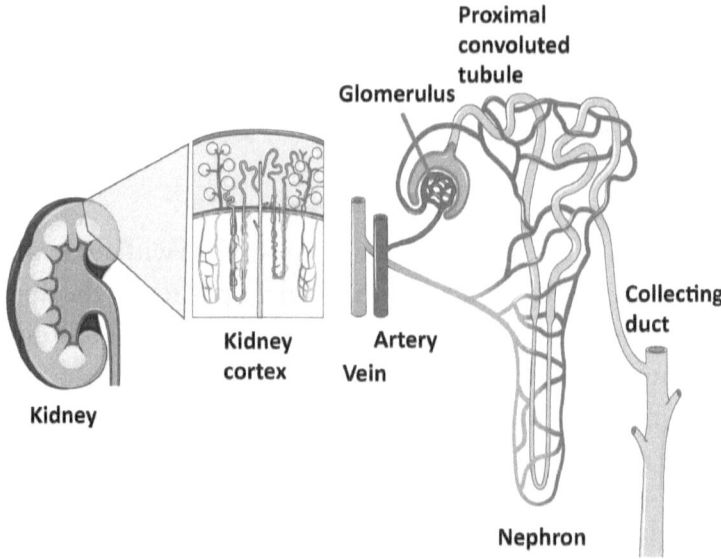

Figure 46. Structure of the kidney and nephron. Servier Medical Art.

How this structure develops in an embryo is nothing short of miraculous. So what do kidneys do? Our body produces several breakdown products that we cannot use for anything. Urea is one of those molecules. It is generated when amino acids, derived from protein digestion, are used as nutrients. Amino acids have a nitrogen atom in their structure. If we want to burn amino acids, we use the Arc de Triomphe cycle, like we do for carbohydrates and fat.

The Arc de Triomphe cycle, which breaks down nutrient fragments as described earlier, contains only carbon atoms (the cyclists), which are rearranged and must donate coins (electrons) to convert nutrient fragments into carbon dioxide. To join the race, the amino acids must leave the nitrogen behind. Our liver is happy to do that and produces urea to get rid of the nitrogen. Another waste product is urate, which is generated when DNA or RNA is broken down. It also contains a lot of nitrogen. Without kidneys, our blood would fill up with nitrogen-containing waste products. They filter the blood and recapture everything valuable including most of the water, and the rest becomes urine.

How does this work? In the first step, we filter the blood in a structure called the glomerulus (Figure 46). It is a collection of blood vessel loops that look like curled sausages. It is a bit like a pressure filter for coffee. The water with extracted colour and flavour gets through the narrow gaps of the blood vessels, but the coffee powder stays behind. The coffee powder would be all the red and white blood cells and larger proteins such as serum albumin and antibodies. The 1.5 million filtration units (glomeruli) filter our blood many times a day, about the equivalent of 180 litres. That would be a lot of urine. We could not drink that much, so we must get it back. Water follows osmotic gradients. If you add water to wilted plants, they become stronger again. This occurs because inside the plant there is water with a lot of minerals and other molecules. Thus, in the same volume of liquid, you find less water molecules and more salt molecules. On the outside, you have just water molecules. As a result, water is more 'concentrated' on the side where it is pure than the inside of the plant where it is diluted by other molecules. The trick is that the plants do not leak the salt, and then the water will flow from concentrated to less concentrated, inflating the leaves. Vice versa, if you bath the same plants in concentrated salt solutions, they will shrivel because water moves in the opposite direction. The kidney's pipes or tubules as they are called are desalting facilities. We all know that blood tastes salty, and when it is filtered in the glomerulus, the salt is in the press juice. The epithelial cells that line the tubules are like our intestinal paddock with kissing gates and farmers. We have a sodium pump (the farmer) on the side facing blood vessels. In Figure 46, the reader can appreciate how close the blood vessels come to the tubules. The cells are not shown but stretch from the tubules close to the blood vessels. On the side where the cells face the lumen of the tubule, there are finger-like extensions like those in intestinal cells. These have numerous embedded kissing gates. I only want to mention two. One kissing gate lets sodium ions in and protons (acid) out. When a sodium ion bounces into the gate, it can pass through the gate; on return, a proton (acid) goes the opposite way. Once inside, the sodium ion will eventually meet the sodium pump which will push it out back into the blood. As a result, salt is removed from the press juice and returned to the blood. The other kissing gate is (almost) the same as the one discovered by Robert K. Crane for glucose and sodium ions in the intestine. Like in the intestine, the farmer is the important part. Without the farmer, there would be no removal of ions and nutrients from the press juice. The juice is generated close to the surface of the kidney. As the press juice goes deeper and deeper into the kidney, the osmolarity[r] of the surrounding tissue becomes higher and higher. This drives the water out of the urine. There is fine-tuning before the pipes merge into the ureter. Eventually the final urine is produced and ends up in the bladder. Remarkably all salt, water, and nutrients are reabsorbed in the few seconds it takes for the press juice to move through the pipes to the core of the kidney. One filtration reabsorption unit is called a nephron.

Marcello Malpighi (Figure 26) discovered the glomeruli in 1666. We met him earlier as the discoverer of blood capillaries. William M. Bowman (1816–1892) then showed in 1842 that the tubules started at the point where blood vessels form little loops inside the glomerulus. The hard shell around these blood vessel loops is still called the Bowman capsule. Carl Ludwig (1816–1895) in 1844 suggested the idea that a press juice is generated in the glomerulus and that the water and nutrients are reabsorbed during passage along the tubules, although he thought it was a passive process. However, his suggestions were disputed, and it took until the twentieth century that the filtration-reabsorption hypothesis was firmly established. Alfred Newton Richards (1876–1966) in the 1920s used very small sharp glass tubes to puncture the glomerulus and to analyse its content. At the point where the filtrate is generated, it contained glucose and chloride ions while the final urine did not. By 1936, Arthur M. Walker and Charles L. Hudson had refined the technique to the point that tubular samples were taken at different distances

---

[r] Osmolarity describes the amount of molecules in a certain volume of a solution. Most of the time this is similar to its salt concentration, but in the kidney also includes organic osmolytes.

from the glomerulus.[125] These showed that glucose was removed early during the passage through the pipes, while chloride was removed in more distal parts. This at the same time demonstrated specific processes for the removal of different ions and nutrients along the nephron. The plant poison phlorizin, which had been known to cause spillover of glucose into urine, was shown to block the removal of glucose during passage through the tubules. It blocks the kissing gate for sodium/glucose pairs. Richards also demonstrated that the bulk of water is removed in the early parts of the tubules. In 1938, it was demonstrated that there is a maximum capacity for glucose transport by the nephron. As blood glucose levels increase in people with diabetes, this threshold is exceeded, and glucose becomes detectable in the urine. It took until 1965 to show that glucose transport in the kidney required sodium ions and therefore followed the same principles as in the intestine.[126]

Joseph von Mering (1849–1908) discovered that the plant toxin phloridizin was able to induce diabetes in animals. This is, however, a pseudo-diabetes because it is caused by blocking the glucose kissing gate in the tubules of the kidney. Glucose spills over into urine because its reabsorption is blocked not because there is too much glucose in blood. The plant toxin was eventually further developed into highly specific inhibitors of the kidney glucose transporter, and these drugs are used today to treat diabetes, because they efficiently remove excess glucose from the body.

8

Hermes Delivers the Message

wanted to see what no one had yet observed,
even if I had to pay for this curiosity with my life.

—Jules Verne

When we consume food, the liver is the first organ to notice its arrival after the intestine. There is a special arm of our circulation called the splanchnic bed. It provides the intestine, stomach, pancreas, spleen, and liver with blood. When arteries branch out into capillaries in the intestine, they collect the nutrients, and from there the path leads straight away to the liver via a blood vessel called the portal vein. This is one of the reasons why in ancient times it was thought that food forms blood and that venous blood comes from the liver and forms tissues. The real reason for this way of plumbing is a different one. With this arrangement, the liver, which is the organ that can metabolise more chemicals than any other organ, immediately notices what is coming into the body. The metabolism of the liver is very much driven by the incoming nutrients, particularly sugars. The liver, with the help of the adipose tissue, kidney, and muscles, ensures that our blood glucose levels are maintained in a close range, which in a healthy person is between 0.75 g and 1 g per litre while fasting, and up to 2 g per litre after a meal. We will talk about its regulation in the next chapter.

Like muscle, liver has a store of glucose in the form of glycogen. Claude Bernard was the first to recognise liver as the organ that provides the body with glucose during fasting. He verified the presence of sugar in the blood of dogs, whether the animals were fed with a diet of sugar, starch, or meat, or given nothing at all for two days.[8] In another experiment, he tied a ligature around the portal vein to prevent nutrients from reaching the liver and found 'not without astonishment' that blood which had flown backward from the liver into the portal vein contained enormous quantities of sugar and concluded 'it was evident that it was the liver from which the sugar arose'. As mentioned, glycogen can provide our body with glucose for up to twelve hours. However, the reservoir can be filled up each time when carbohydrates arrive with a meal. For once the incorporation of new glucose units into glycogen does not require ATP but its cousin nucleotide UTP (uridine triphosphate). Luis Federico Leloir (1906–1987, Figure 47), based on work in yeast and plants, discovered in 1957 that glycogen was synthesised in liver using an unusual nucleotide connected to glucose, namely UTP.

Figure 47. Pioneers of sugar and glycogen metabolism. Left: Luis Federico Leloir. Right: Earl W. Sutherland (Wikimedia Commons).

It looks a bit like Euler's coferment, and it is not clear why most organisms use other nucleotides to chemically activate sugars via these nucleotides instead of ATP. Leloir trained with Frederick Gowland Hopkins (1861–1947, Nobel Prize in 1929), the founder of British biochemistry in Cambridge and briefly worked with Carl and Gerty Cori at Washington University in St. Louis. He generated the main body of his work in Buenos Aires, where he was born. He received the Nobel Prize in Chemistry in 1970 for his discoveries in carbohydrate metabolism.

ATP itself has another critical role in regulating the build-up and breakdown of glycogen, after a meal and when we are fasting, respectively. Let us imagine the glycogen store as our bank account. We also have a wallet for everyday spending (our blood volume). The account is always available, but it does not fill unless we have a payday. Let us be old-fashioned and assume we receive actual cash on payday. The amount of money in the account is regulated by the transfer of money into and out of the account. For the analogy to hold, each meal is a payday, and the time between meals is the time between paydays. After a meal, we use the glucose (money) to fill up our glycogen stores (bank account). When we are fasting between meals, spending goes on, and the bank account empties. In our body, some of the money comes back as if we were withdrawing in large notes and return the change to the bank account. This is the Cori cycle, which we met earlier when we looked at the credit card system of muscle energy generation in chapter 3. Muscle was able to generate ATP very quickly by generating lactate because the blood flow did not provide enough oxygen. That lactate is then returned to the liver where it is recycled into glucose (more on that in a moment), some of which is stored as glycogen. Lactate is also constantly generated by our red blood cells because they do not have any mitochondria. Thus, there is always lactate that can be converted into glucose.

Metabolism in liver is extraordinarily complex. If you draw a map where chemical reactions happen, it looks like Figure 48.

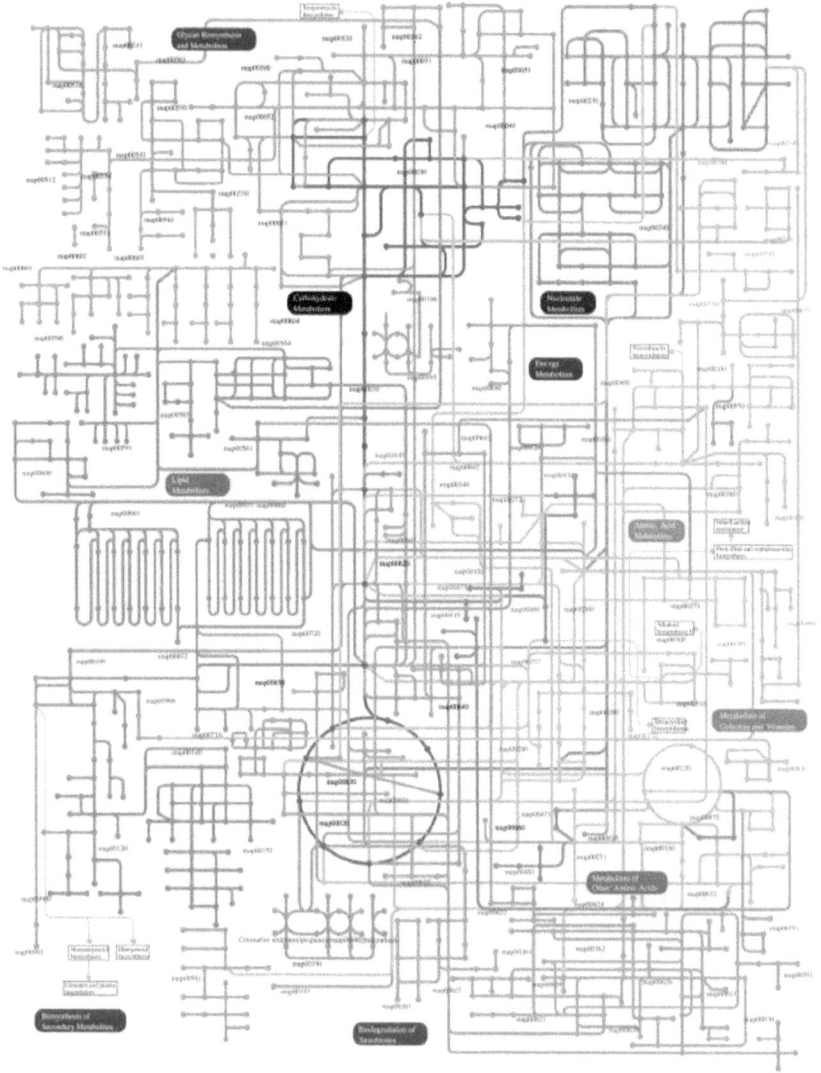

Figure 48. A map of human metabolic pathways. Each dot is a metabolite; each line is a chemical reaction facilitated by an enzyme (Pathways2.embl.de).

In retrospect, it looks a bit silly; but when I studied biochemistry, it was our pride to know all these pathways in detail and be able to reproduce them in exams. Not a chance with students in the age of the internet, but it gives me an appreciation on what is going on in our body. If the reader can spot the Arc de Triomphe cycle on the map, he or she has already learned some biochemistry. A comparison to the map of a large city is pertinent and helpful. It would be useless to have traffic lights at every intersection. Instead, you want to regulate the important intersections. To stay in our analogy, we could see the map as a diagram where our money goes. We can withdraw money or transfer money and use it to pay bills or buy food or consumer goods until it gets refilled. There is more spending when money just came in and less spending at the end of the pay period.

Let us continue with our old-fashioned bank with accountants involved. After a payday, you bring the money to the bank where the money is counted and deposited in your bank account. This is precisely what happens in your body as well. When nutrients are coming in and glucose is recognised, the transfer into the glycogen account is authorised, and it fills up. Spending continues from your wallet, but each time you want to refill your wallet outside paydays, you have to go to the bank and withdraw some money.

How is ATP involved in this? The commands to transfer the money into the account and to withdraw money requires ATP as a trigger. It transfers its phosphate onto the protein that synthesises glycogen or breaks it down. Thus, ATP is the trigger that directs the money into your account or triggers expenditure. The accountant waits for our message that money has arrived. When the receptionist (or receptor) receives a message that money is coming in, he or she calls the accountant to move the money into our account. For spending, the same applies. When we receive an invoice (message), it triggers our response to withdraw money and transfer it to a different account. For regular shopping, we just go to the ATM, withdraw money, and use it for shopping.

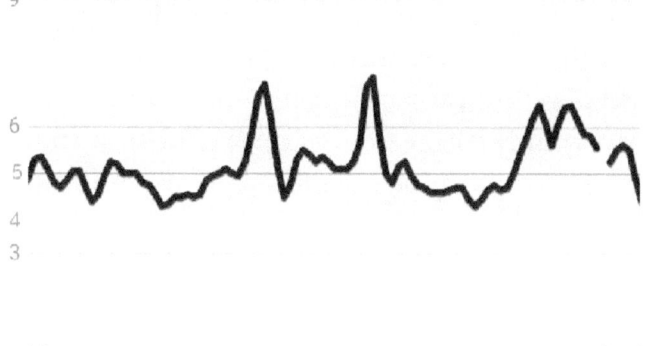

Figure 49. The author's blood glucose readings over a 24 h period. The unit in which glucose was measured is millimole per litre. For comparison, 5 mMol/L equals 0.9 g glucose/L. Breakfast lunch and a longer dinner are readily recognisable.

Figure 49 shows my own blood glucose concentration over a 24 h period, measured with a glucose sensor. It is easy to identify the sharp peaks for breakfast and lunch and a broader peak for dinner, which was a more elaborate meal.

The reason for the sharp drop after breakfast, lunch, and dinner is not that the food suddenly stopped coming in; it is the receptionist receiving the signal that I want to move my money into the bank account. In terms of storing glucose after a meal, our muscles have more capacity than our liver. It is like two bank accounts at two different banks to which we can direct our money.

Carl and Gerty Cori headed the leading laboratory in the 1930s that investigated the deposition and withdrawal of money from the glycogen account. Their laboratory at George Washington University in St. Louis was like Otto Meyerhof's laboratory in Germany, a place where the best minds went to learn biochemistry. At least six scientists who trained with the Coris moved on to become Nobel laureates, namely Arthur Kornberg (1959), Severo Ochoa (1959),

Luis Leloir (1970, Figure 47), Earl W. Sutherland (1971, Figure 47), Christian de Duve (1974), and Edwin Krebs (1992, Figure 50). Earl Sutherland wrote, 'I believe that kind of stimulating environment, with the necessary "critical mass" of young and talented investigators, with the opportunity for the free exchange of ideas, is an important ingredient in the making of scientific progress.'[127] Regarding the glycogen account, Earl W. Sutherland and Edwin G. Krebs played a prominent role because they revealed what triggers deposits and withdrawals and who was the receptionist. Edwin G. Krebs (Figure 50) was joined by Edmond H. Fischer (Figure 50) to work on the withdrawal process.

As we saw in the case of the discovery of ATP, minor changes to experimental procedures can make a significant difference in the outcome of an experiment. Instead of liver, they made extracts from muscle, which as mentioned also stores glycogen. While the Coris had used filter paper to clarify the muscle extracts, Krebs and Fischer used a different method based on fast spinning of samples to remove debris. To their surprise, money withdrawal was not possible after their isolation procedure. Secondly, they found out that they had to use a fresh extract, not an aged one. Earlier in the book, we discussed how distinct types of extracts, such as boiled or fresh, can be used to identify cofactors and coferments. The same was true in this case. Further study revealed that the filter paper bound calcium ions and the factor that disappeared upon aging of the extract was ATP.

By the late 1950s, Krebs and Fischer had established that ATP was required to trigger withdrawal of money (glucose) from the glycogen bank account. Meanwhile, Earl W. Sutherland worked on slices of liver instead of using extracts because he was interested in the receptionist for hormones such as adrenaline and glucagon,[s] for which intact cells held more promise. He found that addition of adrenaline

---

[s]   Glucagon is a hormone released while fasting, which increases blood glucose concentrations.

Figure 50. Edwin G. Krebs (left) and Edmond H. Fischer (right) discovered how proteins can be switched on or off with the help of ATP (Wikimedia Commons).

quickly resulted in money withdrawal.[t] He also established that a phosphate was added to the enzyme (the trigger to withdraw money) when it was activated. This fitted with Krebs and Fischer's observation that ATP was required, which would donate the phosphate.

Further work, using extracts prepared in different ways, showed that a receptionist was involved who transmitted the message that an invoice had arrived. Moreover, the receptionist did not communicate the adrenaline message directly to the bank account but via another message, which is now called a second messenger. This second messenger turned out to be another nucleotide derived from ATP. It is generated by ATP reacting with itself to make an inner circle and is called cyclic AMP.

Figure 51 summarises what happens when we need to mobilise glucose during fasting. Glucagon is the hormone that initiates the process. With the blood, it is delivered as a message to the liver where

---

[t]   Physiologically it is the hormone glucagon that serves as the invoice to release money. At the time, unphysiological concentrations of adrenaline were used, which achieve the same effect. Normally adrenaline causes the breakdown of glycogen in muscle.

it meets the receptionist. The receptionist sends a second messenger on its way, which activates an enzyme (the accountant) that uses ATP to attach a phosphate to the enzyme that breaks down glycogen (releases money). The money is then used for all everyday expenses. These kinds of messenger systems control almost any process in our cells, but it would lead too far to cover all these here. However, we have come across cyclic AMP in the nervous system where it is important for the formation of memory. Nature reuses successful tools to achieve different outcomes in different cell types.

Adrenaline is released during exercise and exerts effects that are like those of glucagon, but its receptionist is found on the surface of muscle cells, not so much in the liver.

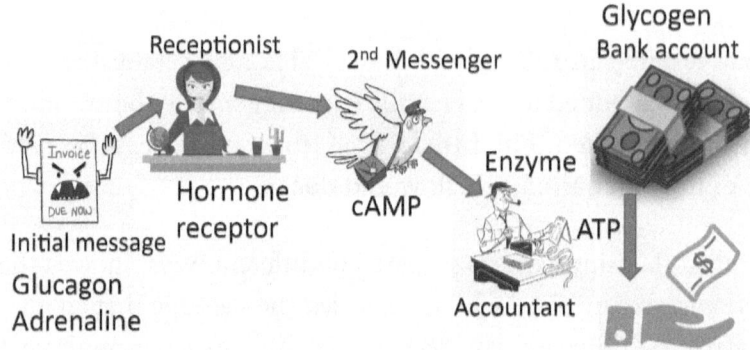

Figure 51. The messenger cascade that provides our body with glucose during fasting. Glucagon is the initial message (invoice) for the liver, while adrenaline has the same action on muscle. The receptionist receives the invoice and sends a second messenger (cyclic AMP) to the accountant who initiates the transfer of money with the help of ATP.

I briefly mentioned that the bank can be open or closed. The liver bank is open 24/7, but muscle is different. Glucose can only enter muscle upon receiving the message that nutrients are coming in. Here insulin is the message, and it tells the muscle bank to open its gates to let glucose (money) in. The muscle bank has quite a few of those gates with roller doors to block them. When money comes in, the roller doors open, allowing the gates to operate. When fasting, the roller doors shut, and the bank is closed.

The person to recognise the roller doors was Polish-borne Rachmiel Levine (1910–1998, Figure 52), who lost both parents and initially emigrated to Canada because he had relatives in the United States. He later moved to Chicago to work on the action of hormones in metabolism. The increased entry of glucose into muscle upon insulin messaging is called the Levine effect and made him 'the father of modern diabetes research'. He had to convince a large community of diabetes researchers that one of the main actions of insulin was to open the roller doors for glucose, not so much changing how fast it was metabolised. In his later years, he encouraged researchers in California to develop the first recombinant insulin, which is now used worldwide to treat type 1 diabetes. More on that in chapter 9.

Figure 52. Pioneers who investigated nutrient metabolism. Left: Rachmiel Levine (National Library of Medicine), Right, Feodor Lynen (Wikimedia Commons).

After establishing how the bank account fills up and empties, we want to investigate what happens when it has run out of money. When glycogen runs out, we must generate glucose from scratch or by gluconeogenesis, as biochemists call it. To stay within the analogy, the liver starts borrowing spare change from other sources to generate money for everyday expenditure. Claude Bernard had already observed in 1848 that dogs fed only lard and tripe had significant amounts of glucose in their blood. Initially surprised, he convinced himself that glucose was generated from other nutrients. He moved on to establish that not only was glucose observed in the blood of diabetic humans, but glucose was always found in blood. These observations made it clear that the liver can generate glucose not only from glycogen. As we saw earlier, Carl and Gerty Cori showed in 1929 that lactate coming from muscle could be converted into glucose and glycogen. They proposed a cycle between muscle and liver in which lactate released from muscle is converted back to glucose in liver, which in turn can be used to fuel muscle or build up its glycogen. We already came across this Cori cycle when we looked at energy production in muscle.[47] At the time, most scientists thought that the generation of glucose from lactate would use the same series of reactions as those used in the breakdown of glucose, the Embden-Meyerhof-Parnas pathway (chapter 2). Herman Kalckar was the first to suggest that certain organs such as the kidney and liver had a specific pathway to generate glucose. It took until 1954 when Merton F. Utter (1917–1980) demonstrated the unique reaction that allowed the conversion of lactate into glucose.[128] In 1963, he found a second unique reaction of the pathway.[129] Both reactions require ATP. In these reactions, animals and humans can incorporate carbon dioxide into organic matter, a type of reaction which is typically thought to occur only in plants. It is only the dominant generation of carbon dioxide from nutrient combustion that eclipses the comparatively small fixation of carbon dioxide. As we saw in chapter 6, lactate is not the only resource to make glucose; we can also use amino acids derived from muscle protein and glycerol derived from adipose tissue.

Now we must look at the other end as well. What happens when there is too much glucose and the bank account is already full? In that case, the liver decides to make fat. It is like exchanging the extra money into gold. That keeps its value but is more difficult to change back into cash.

The chemistry is complex and involves several vitamins, cofactors, and importantly ATP. The series of reactions was worked out by Salih J. Wakil (1927–2019) at the University of Wisconsin and by Feodor Lynen (1911–1979, Figure 52) at the Max Planck Institute for Cellular Chemistry in Munich.

Salih J. Wakil was raised in Iraq and because of his achievements in a pre-university exam received a scholarship to study at the American University of Beirut.[130] He then moved to the United States to do a PhD. His career truly began when he became a research associate at the Institute for Enzyme Research at the University of Wisconsin at Madison, where he identified the reaction sequence of fatty acid synthesis. Feodor Lynen (1911–1979, Figure 52) graduated under Heinrich Otto Wieland (1877–1957, Nobel Prize 1927) who was the first to propose that biological oxidations involve the removal of hydrogens instead of attaching oxygen.[131] The Max Planck Institute for Cellular Chemistry was newly created for Feodor Lynen upon an initiative by Otto Warburg and Otto Hahn (1879–1968). Like the ATP requirement for the generation of glucose, it is also the fixation of carbon dioxide that requires ATP for an early step in the synthesis of fatty acids.

Having explored the role of the liver in maintaining blood glucose concentration, we now want to explore another role of the liver in detoxification.

Ammonia is very toxic to us. If its levels increase too far, the brain starts swelling, which can result in a fatal brain hernia. Yet we must deal with ammonia all the time. Our intestinal microflora, for instance, produces ammonia when breaking down amino acids and related compounds. Even our own body produces ammonia all the time when we use amino acids to generate energy. Because the Arc de Triomphe cycle can only deal with carbon-containing compounds, we must get rid of the nitrogen found in amino acids. By the 1930s, it was known that the liver detoxifies ammonia by converting it into urea. Urea is a harmless organic molecule, which can be eliminated through urine formation. Hans Krebs decided to study the reactions leading to urea formation using tissue slices, a method he had learned in the laboratory of Otto Warburg.[132] There is one amino acid that contains a pre-formed urea molecule, namely arginine, but it was thought at the time that this was an exception and could not explain urea formation from all amino acids. To solve the puzzle of urea formation, Hans Krebs incubated liver slices with many amino acids and was disappointed to see a rather sluggish production of urea except for arginine. Even ammonia alone or in combination with amino acids generated only tiny amounts of urea. In a wider search of nitrogen-containing metabolites on 15 November 1931, his research assistant Kurt Henseleit added an unusual amino acid (called ornithine, in case you ask) to the slices. This amino acid is not one of the pearls that is generated when proteins are digested and therefore was not the immediate focus of investigation. Surprisingly, it together with ammonia generated far more urea. Ammonia or the unusual amino acid alone generated little urea. Thus, an amino acid that is not produced when digesting proteins was the best at stimulating urea synthesis. Even more surprising, the urea was entirely formed from ammonia not using any nitrogen from the stimulating amino acid. When they reduced the amount of the unusual amino acid, they found that one molecule of it could generate twenty molecules of urea if ammonia was present. This led Hans Krebs to consider a cyclic reaction scheme where the unusual amino acid was regenerated. I mentioned above that the amino acid arginine contains a preformed molecule of urea. When urea is split off, it does generate this unusual amino acid, which is eventually used to regenerate arginine.

To explain the whole reaction cycle, we are using cyclists again this time as garbage collectors (Figure 53). Each cyclist can collect two aluminium cans one at a time but disposes them crushed together at a collection point. In this analogy, the cans are ammonia, and the two cans crushed together are urea. The more cyclists we have, the more cans we can collect and dispose of. The cyclists do not contribute anything to the garbage apart from the transport and crushing. This explains why adding the unusual amino acid (cyclists) stimulated the generation of urea (crushed cans) provided ammonia (aluminium cans) was available. In due course, Hans Krebs identified the chemical compound next in the cycle.[132] For the first time, Hans Krebs had shown that a cyclic metabolic process can be used to process other metabolites. I deliberately used cyclists again because we looked at the Arc de Triomphe cycle as a cyclic reaction scheme used to generate carbon dioxide and to remove coins to make ATP. You may remember that this cycle was also worked out by Hans Krebs, but only after he had discovered the urea cycle or Krebs-Henseleit cycle, as it is sometimes called.

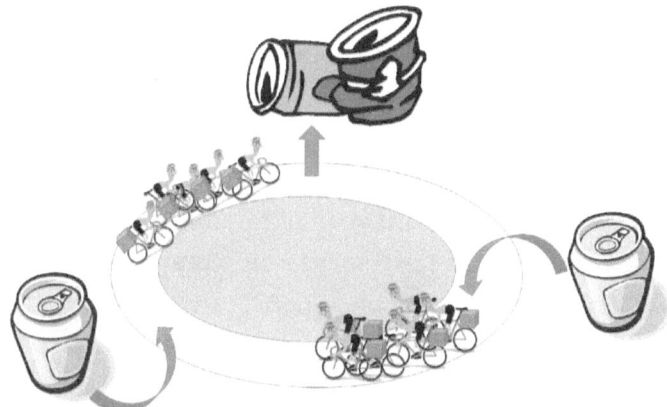

Figure 53. Analogy to explain the urea cycle. Cyclists collect one can of beer at two points, crush them together, and drop them off at the collection centre.

I remember processing liver from the slaughterhouse to demonstrate Krebs's classical experiments to students. It involved a lot of ice-cold acetone, a method which is no longer used in modern biochemistry; and to be honest, the cycle never worked that well apart from the arginine reaction. We should have used slices like Hans Krebs, which preserve the cellular structure, an experimental trick that Krebs realised as important for success.

Franz Knoop (1875–1946), who discovered how fatty acids are metabolised, wrote a letter to Hans Krebs to congratulate him on these findings:[132]

> Dear Doctor Krebs! I thank you very much for your beautiful paper. It is very convincing, how the way in which you come finally to the synthesis of urea by way of arginine and thereby to this significant role of ornithine [the unusual amino acid], which follows entirely by itself out of the sequence of your investigations.

The whole cycle requires energy to run. In total, four molecules of ATP are required to produce one molecule of urea. A costly exercise but essential to protect our brain. The liver is situated in our body to be the first to capture all nutrients coming from the intestine. As we just saw, not only nutrients but also toxic compounds such as ammonia which needs to be defused to urea, come from the intestine. The liver needs to deal with many more toxic compounds and these days also with drugs. We have discussed several ion pumps that energise cells and allow nerve cells to function. There is another group of pumps, also driven by ATP, that protect us from a variety of plant toxins.[133] When we eat plants, we also eat chlorophyll, the green pigment that captures light. A breakdown product of chlorophyll is pheophorbide, which makes skin photosensitive. Unbeknownst to us, pheophorbide is already rejected by the intestine due to the pump that chucks it out the moment it wants to enter our body.[134] Without it, our skin would become brittle when exposed to light. Genetically modified mice without the pump lose the tip of their ears, because they are not covered with fur, get exposed to light, and become brittle.

The intestine and liver have stacks of these so-called multi-resistance pumps.[135] The intestine can reject toxins straight away, while the liver can eliminate molecules it does not want into the bile with which they are released back into the intestine. These pumps also sit in the blood-brain barrier to protect our brains. Ivermectin is a drug widely used in human and veterinary medicine to treat parasite infections. At the same time, it is a neurotoxin and can only be safely used because the walls of brain blood capillaries are furnished with ATP-driven drug pumps. Due to inbreeding, some dog varieties carry a mutation in this drug pump and cannot be treated with ivermectin, because it would accumulate in the brain and cause paralysis.[136] Unfortunately, overexpression of these drug pumps in cancer cells makes them resistant to chemotherapeutics. Because there are several of these pumps, there is not a single drug that can be used to make them sensitive again. What helps us to avoid toxic compounds in food can make renegade cells resistant to drugs, but evolution could not have foreseen medical treatments. More on that in chapter 10.

9

Marjorie Paws the Way

It is a great misfortune to be alone, my friends; and it must be believed that solitude can quickly destroy reason.

—Jules Verne

Our body is particularly good at removing sugar after a meal. To illustrate this, let us calculate the amount of glucose that we would get out of a pizza. A pizza from the freezer section has about 80 g of carbohydrates. The carbs will be digested into glucose because they are mostly starch. The digestion process will increase the weight by 10% because of the water we use to split the starch. An adult male has about 5 litres of blood. Thus, 88 g of digested carbohydrates would translate into 16 g per litre. Yet after a meal, the blood glucose concentration does not exceed 1.5 g per litre if we are healthy. If someone has diabetes, this will be quite a bit higher, perhaps 3 g per litre. While the starches are not digested immediately, it will not take longer than an hour for digestion to be completed. This suggests that glucose is stored somewhere and that this storage does not work as well in diabetes.

ATP is intimately involved in the detection of glucose and is the trigger for the release of insulin when blood glucose concentration rises after a meal. Before we look at the role of ATP, we will review the exciting history of insulin.

There are two forms of diabetes, namely type 1 and type 2. Type 1 is the juvenile diabetes and is an autoimmune disease where the immune system attacks specific cells in the pancreas, called beta cells. Type 2 diabetes is a late-onset form, typically caused by long-term oversupply of food. It is often accompanied by obesity, high blood pressure, and elevated blood cholesterol. The oversupply of nutrients numbs the system to the presence of glucose or more precisely to its messenger insulin. As a result, glucose removal becomes sluggish and blood glucose levels remain high for several hours after a meal and overnight. Elevated levels of glucose in blood plasma result in spillover into the urine because the kidney cannot recapture the glucose completely. Diabetes mellitus literally means 'passing through sweetness'. The disease is also associated with increased urine volume. When looking at the intestine, we discussed that glucose and salt can be used to remove water from the intestine during diarrhoea. The connection between water removal and glucose results in larger urine volumes when glucose is incompletely reabsorbed. The sweet-tasting urine was already known in ancient times but was rediscovered in 1674 by Thomas Willis (1621–1675). He stated that the urine was 'wonderfully sweet as if it was imbued with honey or sugar'.[137] He was referring to type 1 diabetes as type 2 diabetes is a modern affliction. In 1815, the sugar in the blood of diabetic individuals was identified as glucose by Michel Chevreul (1786–1889). Matthew Dobson (1735–1784) recognised that diabetes was a systemic disease and not caused by the kidneys releasing sugar. Claude Bernard, as we discussed earlier, then discovered that glucose could be produced by the liver. The key role of the pancreas in diabetes was discovered by Joseph von Mering (1849–1908) and Oskar Minkowski (1841–1904) who noticed that removal of the pancreas in dogs caused diabetes and that this was unrelated to the digestive enzymes produced by the pancreas (see chapter 7 for the role of the pancreas in food digestion).[138] Paul Langerhans (1847–1888) had earlier described groupings of specialised cells embedded in the normal pancreas tissue. These are still called the islets of Langerhans. Several researchers tried to improve diabetes by injecting pancreas extracts, but the toxicity was too high, or the results were unreliable. At the beginning of the twentieth century, it became clear that it was not the main tissue of the pancreas that was involved in the antidiabetic effect but the islets of Langerhans. It was the Belgian Jean de Meyer (1878–1934) who first suggested the name insulin from the Latin *insula* (island) in 1909 for the substance that controlled glucose metabolism. Edward Albert Sharpey-Schafer (1850–1935) popularised the use of the term *insulin*. The problem of crude pancreas extracts to treat diabetes is the contamination with the digestive enzymes that the pancreas secretes into the intestine. They will destroy the tissue at the site of injection resulting in ulcers and digest insulin. However, Louis Vaillard (1850–1935) and Charles Louis Xavier Arnozan (1852–1928) had discovered that closing of the duct that connects the pancreas to the intestine causes degeneration of the pancreas without affecting the islets of Langerhans and without causing diabetes. Thus, extracts could be made with less contamination.

Figure 54. The discovery of insulin. Left panel: Charles Best (left) and Frederick Banting (right) with Marjorie(?). Right panel: John J. R. MacLeod. (Wikimedia Commons).

This provided the basis for the isolation of insulin, described vividly by science writers Thea Cooper and Arthur Ainsberg.[139] Preparing for a lecture at the University of Western Ontario at the last minute, Frederick Banting (1891–1941, Figure 54) was reading the latest research articles at midnight 31 October 1921. He came across an article describing that a rare pancreatic stone caused degeneration of the pancreas without affecting the islets, thereby confirming older experiments in which the pancreatic duct was tied up. Banting was unaware of earlier attempts to prepare pancreas extracts to improve diabetes, but it also made him unaware of the failures and difficulties. The University of Western Ontario had no laboratory space available, so he went back to Toronto where he had trained in medicine. He approached John James Rickard MacLeod (1876–1935, Figure 54) to provide him with a laboratory and funds. MacLeod assigned a young student to assist Frederick Banting named Charles Best. In their experiments, they paired two dogs. In the first dog, the pancreas duct was tied to induce degeneration. After a couple of days, the complete pancreas was removed from the second dog to induce diabetes. Then pancreas extracts from the first dog were made and injected into the second dog. Despite using extracts from degenerated pancreas, the results remained mixed, most likely because degeneration was incomplete. On 20 August 1921, they were desperate because they did not have enough extract to treat Dog 92. To gather an extract more quickly, they stimulated the release of the digestive enzymes from the pancreas using the hormone secretin and prepared an extract after that. Dog 92 who was almost unresponsive due to diabetes recovered marvellously and could even jump down from her cage without falling. Banting later described the event as one the greatest experiences of his life. Eventually Dog 92 died after twenty-one days because they were running out of extract. The researchers were emotionally devastated. Unreliable extracts remained a problem, and in November 1921, they went to the slaughterhouse to retrieve pancreata from foetal calves. These extracts worked well, and the results were presented at a conference in December 1921 and published in 1922. Using these extracts, the dog Marjorie survived for seventy days without a pancreas. In the audience of the conference was George Henry Alexander 'Alec' Clowes, the research director of Eli Lilly and Company. Alec Clowes approached Banting and offered the help of the company to scale up the production of extracts and to improve their efficacy. However, J. J. R. MacLeod, who co-authored the meeting paper, declined the offer and instead initiated production of insulin at the University of Toronto.[139] Because the extracts were too crude for injection into humans, MacLeod asked biochemist James Bertram Collip to help with extraction. He devised a method of extraction with increasing concentrations of alcohol, which yielded a much purer insulin preparation. On 23 January 1922, his insulin preparation was injected into Leonard Thompson upon which the spillover of glucose into his urine disappeared, indicating that his blood glucose had dropped to normal levels. Frederick Banting, irate at the best of times, did not appreciate the progress made by Collip, even attacking him. Had he known that he would share the Nobel Prize in 1923 with MacLeod, he might have been more mellow. Because he fell out with MacLeod and Collip, Banting lost interest in the project and stopped working on the project for a while, although he was receiving attractive offers. Harvey Kellogg, for instance, offered to build Banting his own wing at the Battle Creek Sanatorium and to pay him an annual salary of ten thousand dollars. During his research, Banting had received no compensation and frequented public soup kitchens for sustenance. Eventually he returned to the clinic in Toronto though.

In the meantime, the University of Toronto still struggled to produce insulin preparations of reliable quality, and eventually it was agreed to involve Eli Lilly and Company. At that point, insulin was already in high demand; but even for the most serious cases, not enough insulin was available. By August 1922, Eli Lilly was able to produce small amounts of insulin on a regular schedule; and in 1923, production increased to the point that an estimated 7,500 physicians could prescribe insulin to 25,000 patients. Sales exceeded $1 million.

Before insulin, diabetics would eventually fall into a coma induced by highly elevated levels of glucose. One of the very few life-extending measures available before insulin was adhering to a very low-calorie diet. Elisabeth Hughes, daughter of the United States secretary of state, was the first celebrity patient to receive insulin in 1922; and her recovery made the news. She had been on a near-starvation diet for over three years and would write,[139]

> Nobody up here except Dr. Banting of course knows what a big diet I'am on and the foods I'am eating on it either, and I know if they did know they'd nearly roll off their seats. It's our big secret! I could use up pages just in innumerating all the dishes I have nowadays and it seems to me that I eat something everyday that I haven't tasted for over three years, and you don't know how good it seems and how I appreciate every morsel I eat.

She gained 10 kg of weight over a couple of weeks and 25 kg in half a year.

Insulin is our main hormone to reduce blood glucose levels. As mentioned in the previous chapter, it does so by opening the roller doors to the banks in muscle and adipose tissue, which then can accept money (glucose) into the savings account. In addition, the liver glycogen bank account also saves money. Altogether this removes glucose quite efficiently, storing it as glycogen and fat. In people with type 1 diabetes, the body is not capable of storing glucose. As a result, the body is starving although nutrients are available. Injecting insulin restores the ability to store glucose, resulting in weight gain and normalisation of metabolism.

In this chapter, we also want to look at the mechanism by which our body detects glucose and orders the release of insulin because it involves ATP. In the 1930s, it was already established that the elevated levels of blood glucose stimulate the secretion of insulin.[140] By the 1960s, it had been shown that glucose had to be metabolised for insulin to be released. It was also known that insulin was stored in small packages. This is like the nervous system, where an incoming drop of the voltage caused the release of packages of neurotransmitters. In further similarity, it was established in 1968 that the beta cells of the pancreas changed their membrane voltage in response to glucose.[141] There are several ways to change the membrane voltage. In an electrical power grid, the voltage will increase when too much power is produced and not enough power is consumed. On the other hand, if too much power is consumed and power production is not raised, the grid collapses. This happens occasionally when during the peak of summer, too much power is used for air conditioning. In a similar way, beta cells can change their grid voltage. They reduce the power output, and because the cellular consumers remain the same, the voltage collapses. The mechanism that keeps the battery charged in beta cells is the movement of ions across the cell membrane, in this case potassium ions (Figure 55). This was demonstrated in 1978 by Jean-Claude Henquin .[142] In 1984, Frances M. Ashcroft of Oxford University showed that incubation with glucose stopped the release of potassium ions through a specific pore, thereby destabilising the grid voltage. A year later, P. Rorsman and G. Trube demonstrated that these gates are regulated by ATP.[143] The gene that encodes the ATP-regulated potassium pore was eventually identified in 1995 by Lydia Aguilar-Bryan and colleagues.

These results gave rise to the fuel hypothesis that stated that the release of insulin was directly correlated to the amount of fuel present in blood, which in turn is correlated to the amount of ATP produced.[144] ADP opposes the action of ATP so that the channel carefully monitors the value of the ATP currency at any time.[145] In chapter 1, I mentioned that our body is desperate to keep the value of ATP close to one dollar, and any decline of its value indicates an energy crisis. Beta cells are the exception because they are designed to measure the energy or fuel content of our blood.[145] As we saw earlier, muscle cells monitor the amount of AMP to detect energy fatigue, but the response to fatigue is slow and results in the mobilisation of energy reserves such as fatty acids. The beta cell needs to respond more quickly to reverse the sharp increase of glucose in blood after a meal (Figure 49) and directly monitors the ratio between ATP and ADP, which is the value of ATP.

When type 2 diabetes sets in, glucose is no longer efficiently removed by muscle, adipose tissue, and liver. The body becomes insulin resistant. Initially this can be compensated by enhancing the release of insulin, which can be achieved by a group of drugs called sulphonylureas. Their discovery is another intriguing story.

In 1940, sulphonamides had been discovered as the first antibiotics.[146] In 1942, Marcel Janbon at the hospital in Montpellier evaluated a new sulphonamide for the treatment of typhoid fever.[147] After the unexplained death of some of their patients, Janbon realised that it had caused dangerously low levels of blood glucose. Over the next couple of years, Auguste Loubatiere (1912–1977) then undertook more careful studies of its glucose-lowering actions and found out that the drug did not work when the pancreas had been removed in dogs. Moreover, very small doses were effective in reducing blood glucose when the drug was injected directly into the artery supplying the pancreas.

Similar side effects of sulphonamides were observed by Franke and Fuchs in Berlin in 1954. Carbutamide, as the compound was known, was then tested in diabetic subjects who did not require insulin treatment. Shortly after, a new compound, the sulphonylurea tolbutamide, was synthesised which had no antibacterial action but retained its action on blood glucose levels. By the end of the 1950s, it had been established that sulphonylureas increased the release of insulin. It took until the 1980s until it was demonstrated that sulphonylurea compounds blocked the very same potassium release pores as ATP. They do not occupy the same spot but have similar actions.

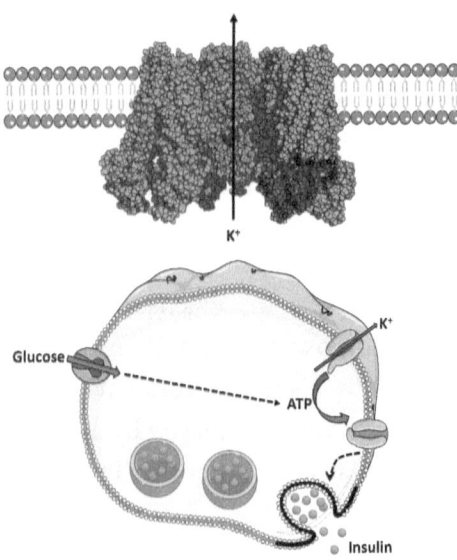

Figure 55. Upper panel, the pore that releases potassium ions from beta cells. ATP blocks the channel. Lower panel, blocking the potassium channel triggers the release of insulin. This is caused by a collapse of the cellular membrane voltage.

Insulin had its fair share of Nobel Prizes after Banting and MacLeod. Dorothy Hodgkin (1910–1994, Figure 56) was born in Cairo before the First World War.[148] When the war began, her mother brought her back to Britain where she was raised by a nurse and older relatives. Dorothy Hodgkin often attributed her independent spirit to this period in her childhood. In 1928, she started studying chemistry at Oxford University. She became fascinated by crystals and worked for her doctorate with John Desmond Bernal (1901–1971), who used x-ray techniques to study biological molecules. In 1934, they received crystals of pepsin and learned that crystals of biological molecules had to remain moist to work with them. She went on to elucidate the structure of penicillin and of vitamin B_{12} before attempting insulin. In 1935, she succeeded in producing insulin crystals, remembering it as the most exciting moment in her life. As we discussed, for the structure of haemoglobin, the mathematical tools to calculate a structure from the x-ray diffraction pattern were not available at the time. In 1955, Frederick Sanger (1918–2013, Figure 56) succeeded in determining the order of pearls on the insulin necklace by using sophisticated chemical reactions. This helped to resolve the final structure of insulin which was announced in 1969. Dorothy Hodgkin received the Nobel Prize in 1964 for her overall work on the structure of biomolecules. Frederick Sanger received the Nobel Prize in 1958 for determining the order of pearls of the insulin molecule and another in 1980 for designing a method to determine the sequence of any DNA molecule.

Figure 56. Pioneers of insulin structure: Left, Dorothy Hodgkin and right, Frederick Sanger (Wikimedia Commons).

Insulin created more drama over the years, most notably when biotechnology had progressed to the point where production of recombinant human insulin was becoming feasible. The story has been told several times and is summarised here.[149-151] Insulin was selected because its medical application was obvious, and it was such a famous protein. The process involves ATP as a pearl to generate the gene together with its cousins TTP, GTP, and CTP and also as an energy provider to fuse pieces of DNA together. Three groups would compete for the achievement to express human insulin in bacteria for the first time: Harvard professor Walter Gilbert and his team, Herbert Boyer and his team in San Francisco, and William Rutter and his team also in San Francisco.

No laboratory at the forefront of science works without postdoctoral scientists who want to get a high-profile publication from their work in a top laboratory. Walter Gilbert was joined by Argiris Efstratiadis, William Rutter attracted John Chirgwin and German postdoc Axel Ullrich, and Herbert Boyer employed Herb Heyneker. All groups started working on insulin in 1976. Progress in Harvard was hampered by strict biosafety regulations. Because nobody knew at the time whether producing human genes in bacteria could have unforeseen consequences, high-level containment was required to carry out the work. While Harvard was building the appropriate laboratories, a public debate erupted which resulted in a moratorium for work involving cloning of human genes until 1977. Meanwhile, other technical problems beset the Rutter team. Isolation of the coding message, initially from rat pancreata, turned out to be exceedingly difficult because the digestive enzymes of the pancreas destroyed the coding message whenever the tissue was macerated for processing. Eventually Chirgwin devised a method that inactivated the enzymes without harming the coding message at the end of 1976. This allowed the cloning of the rat insulin gene in early 1977, but it was in a form that would not allow production of insulin. It was a successful test run of the methods required to isolate the human insulin gene. However, the group could not publish their achievements because the tools they used at the time were not yet approved for research by the National Institute of Health. Even worse, they were ordered to destroy the bacteria containing the insulin gene, but they did keep the original coding gene joined to bacterial DNA in isolated form. Eventually they had to join the insulin gene to an approved bacterial DNA and published the isolation of the rat insulin gene in May 1977. Because of the commercial implications, Herbert Boyer had joined forces with businessperson Robert Swanson in 1976 to found Genentech and to raise money from venture capitalists. While the Boyer team was behind in time, they had a big advantage. Because the insulin gene is quite small, it was feasible to chemically synthesise its DNA adding pearl by pearl. This circumvented any regulatory hurdles which applied to human DNA. For the synthesis of the human insulin DNA, they joined forces with Arthur Riggs and Keiichi Itakura from Los Angeles. Nevertheless, Riggs and Itakura wanted to synthesise a smaller hormone as a test case before attempting insulin. In December 1977, the team announced the production of the hormone somatostatin in bacterial cells.

Walter Gilbert recruited another postdoctoral fellow, Lydia Villa-Komaroff, who succeeded together with Argiris Efstratiadis in joining the rat insulin gene to bacterial DNA by mid-1977. Gilbert filed a patent, which soon would be licenced to the firm he was founding, Biogen. To attempt the cloning of the human insulin gene, they obtained material from a rare tumour that arises from beta cells in the pancreas, an insulinoma. To carry out the procedure, they had to use facilities of the highest biosecurity, which at the time were only available in germ warfare facilities. For this, the team had to move to Porton Down in Britain. Tragically, some of the material used must have been contaminated because they isolated the rat insulin again instead of the human gene and had to come back empty-handed. The Rutter group tried the same strategy and sent Axel Ullrich to Strasbourg to isolate and insert the human insulin gene into bacteria in mid-1978. At this point, Eli Lilly signed a research contract with the University of California in San Francisco to finance Rutter's team to achieve the cloning and bacterial production of the human insulin gene.

After the success of somatostatin, Bob Swanson could raise more money for Genentech, and the company started its own laboratory facilities. They recruited Dennis Kleid and his postdoctoral researcher David Goeddel. At the same time, Keiichi Itakura's team synthesised the human insulin gene in two pieces,[u] which were separately joined with bacterial DNA by David Goeddel and Dennis Kleid. The

---

[u]   Initially a precursor of insulin is produced from a single messenger RNA. The resulting insulin precursor is processed and cut into two pieces, which remain together, to generate the final insulin molecule. The process, which bacteria cannot do, can be bypassed by producing the two pieces separately.

production of the two parts of human insulin was also done separately before joining the two resulting strings of pearls into the final insulin molecule. On 6 September 1978, Genentech announced that they had successfully produced insulin in bacteria and therefore started a new era in biotechnology. Eli Lilly instantly signed a contract with Genentech to be part of the insulin production. Axel Ullrich returned from Strasbourg to join Genentech, and the Gilbert team gave up cloning insulin.

What has ATP got to do with this? The energy inherent to the phosphate groups attached to the nucleotides is used to join the DNA molecules before introducing them into bacteria. ATP is added to the reaction, and with the appropriate enzyme, the energy is used to join two DNA molecules. Moreover, it is not only ATP, but also GTP, CTP, and TTP (the deoxy variants in DNA though) that bring along their own energy when DNA is synthesised by attaching nucleotide after nucleotide. In the process, two phosphates are lost, and the remaining phosphate becomes part of the rungs that make up the twisted DNA ladder.

10

The Hydrophobic Vacuum Cleaner

We may brave human laws, but we cannot resist natural ones.
—Jules Verne, 20,000 Leagues Under the Sea

Thus far, we have looked at the role of ATP in cells and organs under physiological conditions. I want to spend a bit of time looking at the role of ATP in cancer. It is a diagnosis everybody dreads, yet considerable progress has been made in the treatment of a variety of cancer subtypes, and the hope is that eventually we will have tailored therapies for all distinct types of cancer. If you had hope that we could deplete cancer cells of ATP as an energy source, this is unlikely to work for two reasons. Firstly, cancer cells have a generous safety margin when it comes to generating ATP; they are rather limited by generating building blocks to make a new cell, not by the energy that it requires. Secondly, other cells like neurons and muscle cells in your heart are much more sensitive to depletion of ATP. How would you create a safety margin, or therapeutic window as it is called, for a drug that compromises energy production? In fact, trials to slow down the breakdown of glucose have failed to reduce cancer cell growth. Cancer cells have a peculiar way to generate ATP; they mostly use glucose but then largely bypass their mitochondria although oxygen is present. The first person to note this behaviour was Otto Warburg.[152] We have met Otto Warburg in chapter 3 as the discoverer of the enzyme that reacts with oxygen to form water and the discoverer of coferments that store electrons.

The bypassing of mitochondria for energy production from glucose allows them to provide more building blocks, while the speed of glucose breakdown is very much regulated by ATP demand.

However, ATP and its cousin GTP have a far more delicate role in cancer cells. Before we can discuss this role, we need to appreciate the genetic basis of cancer and what discriminates a cancer cell from a normal cell. All our cells have checks and balances to avoid that a tissue cell suddenly starts dividing and forming non-functional tissue. However, some of our organs can regenerate. For instance, if a lobe of the liver is removed, the liver will grow back to its original size but not any further. The epithelial cells of the intestine continuously turn over being replaced by younger cells that move up the villi. Wound healing is another example where skin cells divide and form scar tissue. Other organs do not regenerate, for instance the heart, brain, and kidney. As a result, controlled cell division is a normal occurrence in our body. It is halted by cell-to-cell contacts, which prevent further growth. Apart from immune cells, tissue cells do not move away from their origin.

Why are cancer cells different? The genome is the same in all our cells, but the genes that are active in a certain cell type differ from organ to organ. This is caused by small modifications of the DNA and how the DNA is packaged in each cell type, which changes its accessibility. During embryo development, cells divide and move all the time. However, at this stage, the cells are considered pluripotent, which means they can undergo different fates and develop into different final mature cell types. Embryonic stem cells, which start growing after fertilisation of an egg, can even develop into any cell type. As a result, they are called omnipotent [153]. When cells mature and differentiate, part of the genome is silenced and buried, and only the regions of the genome required for the functions of the tissue are active. Even within tissues, further specialisation occurs. This prevents any wrongdoing once cells have settled down. It is thought that cancer cells revert to a more pluripotent state in which they forget their specific tissue origin, can migrate into other tissues, and grow. But how is this triggered? During our lifetime, we are exposed to different hazards that can mutate our DNA, such as cosmic radiation, UV radiation, and toxic compounds. Normally this is repaired but not all the time. As a result, we accumulate mutations in the cells of our body, but in each cell, the mutations are different because it is a random process. Cell growth and division is controlled by specific hormones called growth factors, which are secreted to initiate cell division, for instance in wound healing or when liver regenerates. When growth factors bind to the receptionist, the message is delivered further, and sometimes multiple copies of the message are given to different sites. The final delivery address inside the cell is a group of enzymes called protein kinases, which require ATP. We have encountered them before in chapter 8 where they initiated transfer of money into the savings account. To do so, ATP was used by these enzymes to fix a phosphate on a target protein, which acted as a trigger to deposit money or pay an invoice. In the liver, this regulates the synthesis and breakdown of glycogen. For cell growth, several of these switches are involved operating like a relay. Enzyme 1 puts a phosphate on Enzyme 2, which gets activated and puts a phosphate on Enzyme 3. The final targets are proteins that alter the activation of large groups of genes involved in cell proliferation. These proteins normally reside in the main part of the cell, called the cytoplasm. Attaching a phosphate can be a signal to move them to the cell nucleus where all the DNA is and where they can start activating specific genes. It turns out that in many cancers, spontaneous mutations have arisen that activate such a relay without a message from the receptionist. One of those mutations is rarely enough to convert a cell into a cancer cell. Normally several mutations must be accrued, for instance in proteins that suppress cell division or in genes that repair damage to DNA. Once enough mutations have arisen, groups of genes start to get activated that would normally be silent.

This explains three observations. Firstly, cancer is mostly a disease of the elderly. We need to accumulate mutations over our lifetime before something happens. Secondly, the process is random; different people develop different cancers. Thirdly, people can be predisposed to a particular kind of cancer if they have inherited a mutation from their parents. Most mutations occur in cells that are not germ cells, so the mutations are not inherited to the offspring. However, if this happens, the next generation starts already with one problematic mutation. Breast cancer is a classical example, where a predisposing mutation strongly increases the risk for it to occur. The promise of personalised medicine is that reading of the genome from a cancer biopsy will tell the oncologist what drugs to use for optimal treatment.

One successful treatment example is a cancer called chronic myelogenous leukaemia (CML).[154,155] This rare cancer is caused by a rearrangement of chromosomes, known as the Philadelphia chromosome. In this rearrangement, a fragment of chromosome[v] 9 exchanges places with a fragment of chromosome 22. This is one of the more complex genome mutations that can happen and is normally prevented. Even more bizarrely the break and merge bring together the halves of two unrelated proteins, generating a chimeric[w] protein which misbehaves. It generates a phosphate-transferring protein that is always active instead of being switchable. The two proteins are called BCR and ABL, and the chimera is hence BCR-ABL. This drives proliferation in a particular type of blood cell called granulocytes, resulting in this type of leukaemia. If the protein can be switched off, the cells would become normal again. In 2009, the

---

[v] The genetic instructions for a functioning cell are not in one continuous string of DNA but rather in forty-six strings of different length called chromosomes.

[w] A chimeric protein is made up of two different halves derived from separate proteins.

Lasker-DeBakey Clinical Medical Research Award was given to Brian Druker, Nicholas Lydon, and Charles Sawyer (Figure 57) for the development of a new drug, called Gleevec, that blocks the access of ATP to the BCR-ABL protein.

Figure 57. Developers of Gleevec, an anticancer drug that targets an ATP-dependent protein controlling cell growth. From left: Brian Druker, Nicholas Lydon, Charles Sawyers (Wikimedia Commons).

It has converted a fatal cancer into a manageable chronic condition, but the medication must be taken forever. Although all phosphate-transferring proteins have ATP binding sites, they seem to be different enough to allow the generation of specific blockers. The development of this drug demonstrated two important advances: first, the promise of personalised medicine in which a known genetic defect can be treated with a tailored medication; second, the possibility to hold cancer at bay with a drug that interferes with a major driver of cellular proliferation.

However, in most cases, this is far more difficult. In 1982, Robert Weinberg and colleagues discovered that the RAS gene was mutated and activated in certain types of cancers.[154] Subsequently, RAS mutations were identified in many more cancers. It is thought that RAS is one of the major cancer-causing mutations in 85% of cancers. When RAS gets the call from the receptionist, it does not use ATP to add a phosphate to another protein. Instead, it is activated by its cousin GTP, which is used similarly to ATP for several cellular functions. However, RAS only needs to bind GTP to become active. Quite the opposite, when it splits GTP to GDP, it puts an end to the activation (Figure 58).

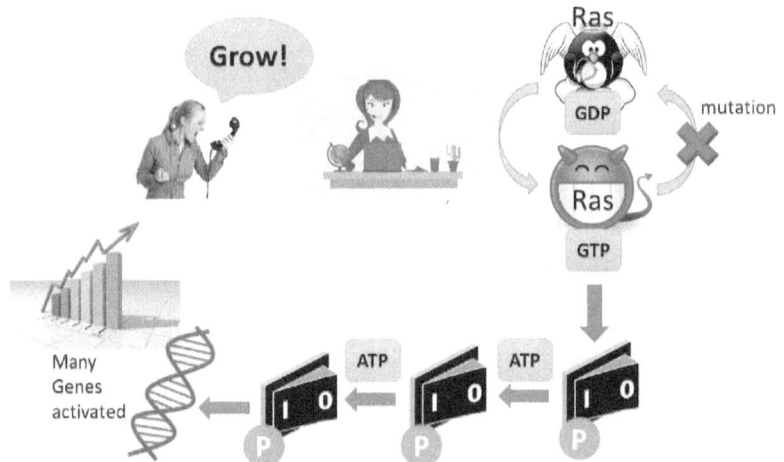

Figure 58. Constant activation of cell proliferation by mutations in the RAS gene. The mutated RAS does not inactivate because it cannot split GTP. The subsequent cascade, which involves many proteins with ATP switches, is permanently switched on resulting in cell proliferation.

Then GDP is exchanged for another GTP, because there is much more GTP than GDP, and the whole cycle starts again. The splitting process is slow, so the kinase remains active for a short while. The GTP bound form of RAS then recruits another phosphate transferring enzyme, thereby transmitting the signal. What is required to turn this signal on and off can become a trap. If a mutation occurs that renders RAS incapable of splitting GTP, it is locked into the 'on' position. In general, mutations are far more likely to destroy function than to create something new and useful. It is like a modern car – try to change a part yourself, and it is more likely to mess things up than to improve the function of the car. RAS sits right at the beginning of a messenger relay that entails several phosphate-transferring enzymes that use ATP to switch other proteins on. Eventually this results in the activation of many genes that initiate and promote cell proliferation. If RAS is permanently on, the follow-on signals are also permanently on. Again, it is not the only mutation in a malignant cancer because there are checks and balances; but in combination with a handful additional mutations, the cancer cell starts growing, pushing aside functional cells.

Mutated RAS presents an odd situation for the development of drugs. Most drugs block an enzyme or block a receptionist, but in this case, the RAS protein is already blocked by the mutation and permanently so. In a car, you exchange the faulty part with a functional replacement. That is the promise of gene therapy but not a quick solution for a disease. GTP binds strongly to RAS and is not easily pushed aside by a drug. This is where thorough research can find ways to bypass the problem. New drugs try to change the ratio between GTP and GDP binding, increase the natural recycling of the RAS protein, or prevent it from docking to the following messengers.[156]

'The' cancer cure does not exist, but the future will bring new tailored treatments for specific cancers based on the mutations that drive its expansion. There is however one more big obstacle that involves ATP as well. In the previous chapter, we learned that ATP-driven pumps in the intestine and the blood-brain barrier protect us against plant toxins. These plant toxins are chemically diverse, and as a result, the defensive pumps recognise a wide range of chemicals. Unfortunately, this includes drugs designed to treat cancers. In fact, quite a few cancer chemotherapeutics are plant products. It is a common observation in the treatment of cancers that the first round of chemotherapeutics is tolerated quite well, and the effect is nothing short of miraculous. The cancer shrinks and is no longer detectable, and the patient seems to be fine. There is a good reason cancer therapy is followed up by diagnostic screens for five sometimes ten years. All of a sudden the cancer pops up again. Some cells have survived somewhere in the body. More importantly, the survivors were selected from a large group of initial cancer cells because they were a bit more resistant to the chemotherapy because they accidentally had more defensive pumps. When the chemotherapy finished, they remained alive and slowly grew up until the diagnostic screen picks them up again. Two things have happened. Firstly, the cells have moved around and have invaded other tissues, and secondly, they are more resistant to chemotherapy because the chemotherapy selected for cells that have more pumps to chuck out cancer drugs. A new round of chemotherapy is initiated. This time successful again, but not as complete as in the beginning. The cycle is repeated, and each time the success is less. What has been an ingenious invention to protect us against plant toxins in food turns into a major obstacle when we want to treat cancer. Even Gleevec, the wonder drug, fades out over time. This time it is not lacking in research; the multiple drug resistance pumps have been studied for decades. However, there is not one but several of them.

The two main ones are called P-glycoprotein (P-gp) and multidrug resistance protein 1 (MRP1).[157] The P-gp protein was identified by Victor Ling and Jack Riordan in Toronto and the gene subsequently isolated by several groups in 1986. The gene for MRP1 was identified in 1992 by Susan Cole.

The role of ATP was established after observing that blocking of the ATP production in isolated drug-resistant cancer cells, increased the uptake of several drugs more than a hundredfold.[158] The astonishing variety of compounds being removed by P-gp has been explained by the 'hydrophobic vacuum cleaner' model. Hydrophobic (water averse) indicates that many drugs are not very water soluble. This is important so they can cross the soap bubble barrier (membrane) surrounding cells. While in the barrier, the pump can pick them up and chuck them out again. As a result, the pump can pick up any compound that can immerse itself in the soap bubble barrier. The only thing it needs to discriminate against is the bubble material itself called membrane lipids. Evolution has found a way to optimise a drug pump selective enough to avoid picking up lipids from the membrane but still selective enough to remove any other compound that penetrates this environment.

Meanwhile, three generations of compounds have been developed to block P-pg and MRP1 [159]. Because the pumps also protect the brain, toxicity can be problem. Moreover, combination therapy is required, but the regulatory hurdles to make clinical trials with mixed unknown compounds are incredibly high.

Cancer treatments still have a long way to go, but each year there is progress for one or the other treatment scheme, and many involve ATP using proteins.

11

The Demon Under the Microscope

The sea is everything. It covers seven tenths of the terrestrial globe.
Its breath is pure and healthy. It is an immense desert, where
man is never lonely, for he feels life stirring on all sides. The sea is
only the embodiment of a supernatural and wonderful existence.
It is nothing but love and emotion; it is the Living Infinite.
—Jules Verne, *20,000 Leagues Under the Sea*

Put two ships in the open sea, without wind or tide, and, at last they
will come together. Throw two planets into space, and they will fall
one on the other. Place two enemies in the midst of a crowd, and they
will inevitably meet; it is a fatality, a question of time; that is all.
—Jules Verne

What is life? This is one of the fascinating questions that has challenged many scientists. The question is intricately linked to the question how has life emerged on earth? We even struggle defining 'life'. Wikipedia says, 'There is currently no consensus regarding the definition of life. One popular definition is that organisms are open systems that maintain homeostasis, are composed of cells, have a life cycle, undergo metabolism, can grow, adapt to their environment, respond to stimuli, reproduce, and evolve.' Erwin Schrödinger, who developed the equations that explain the reactivity of molecules, was one of the first scientists to make a serious attempt at the topic in 1944. His book *What Is Life* inspired many scientists in the second half of the twentieth century to move into life science as the new frontier after quantum theory, notably James Watson and Francis Crick. I am certainly not in the league of Erwin Schrödinger or other Nobel laureates, but this is my opportunity to have a go at this topic.

On this planet at least, everything alive has ATP. As a biochemist, I would propose that any system that autonomously maintains ATP, while using it for work, is alive. *Autonomously* is the key word here. In a famous experiment, Efraim Racker made a small experimental model system which contained a light-driven proton (acid) pump and the radial engine that makes ATP. The reader may remember that ATP is made during the rotations of the engine and that it is driven by a proton battery. When he directed a beam of light onto the system, it made ATP.[160] This proved Peter Mitchell's concept of ATP production, which stipulated that the proton battery and radial engine alone are sufficient to make ATP. However, this is not life because Racker had to isolate the different components and incorporate them into membranes. Moreover, he had to add all chemicals to the system, which he purchased. As a result, the system was not autonomous.

To understand the difference between artificial systems and life, I find it instructive to put a cell under an exceptionally large microscope. Such a microscope does not exist yet, but for the sake of the comparison, we want to enlarge a cell a billion-fold. That is a lot, because it is like stretching a 38 cm ruler to reach the moon. On this scale, a normal cell in our body would have a diameter of 10–100 km. We are talking about a large city. The city is old-fashioned and has a wall all around it, about 5 m wide. You can now appreciate why I compared the cell membrane to a soap bubble earlier in the book. Compared to the city, it is thin, and you must imagine the membrane extending in all three dimensions without collapsing. Well, bursting, because it is all filled with water; and if the sodium pump would not push out the sodium ions all the time, it would burst. In three dimensions the analogy becomes a bit tedious, so for the purpose of the comparison, we are looking at a thin slice of the cell to make it flat like a real city. The library has its own quarter, which is 6 km in diameter. It has its own wall but with large gates to allow traffic in and out. The library is truly bizarre as the letters are arranged in large sewer pipes of about 2 m in diameter. This is of course our DNA. Each letter is almost a metre in length, so just two letters fit across the pipe, and about 10 are stacked up in 3 m pipe length. But the pipes are everywhere, some tightly curled up, some forming large loops. You must imagine the pipes as flexible, and there is heavy machinery working on the loops generating copies of the letters in the tubes. The copies are like a long human chain comprising several thousand people holding hands, each person corresponding to a letter. These copies are called messenger ribonucleic acids (mRNA). The whole human chain gets guided out of the library into the main part of the city. Here they give instructions to gigantic robots to assemble cars and trucks. A group of three humans tells the robots to add a certain part of the car or truck that is assembled. To translate the analogy, the robots are called ribosomes and the cars and trucks are proteins. The city is very energy-hungry because it has more than a handful of power stations scattered around the city. Like in real life, power stations are quite large at 0.5–1 km across. There are a lot of cars and trucks in the city; they represent the proteins. The traffic is frantic. You must imagine peak commuting time, truck behind truck, cars in between and no rules, no roads, and no signs. Large cities in India are an example of order and emptiness by comparison. Everything is tarmacked so trucks and cars will just drive around randomly. But this is not all. Nutrients like sugar and the breakdown products that we find, for instance in the Arc de Triomphe cycle, would be the size of small persons or animals, such as cats and dogs. Imagine that between the trucks and cars, everything is packed with people, dogs, and cats. Before you get nervous, I can tell you that nobody gets hurt. Although cars, trucks, people, dogs, and cats are constantly bumping into each other, they are all unbreakable. However, it is a Harry Potter kind of town. When a person bumps into the right truck, it may be converted into two dogs, or a dog gets rearranged into a cat. To drive the craziness further, a water molecule at this magnification is the size of a soccer ball. Thus, we need to cover the whole city in soccer balls, but they should be translucent and, if anything, facilitate the movement of trucks and people. At this stage, you still imagine the city as too sleepy. While the trucks move around at a pace that you would expect during peak hour, people, dogs, and cats are whizzing around at 300–600 km/h, constantly bumping into trucks and cars. The water balls are even faster, flying around at the speed of sound. This is another Harry Potter magic occurring when you look at things in a gigantic microscope.

Now we must touch on ATP, of course.

In the power stations, ATP is produced by the radial engine, which is a 20 m tall structure. The size of ATP would be that of a large person. These people can leave the power station and help with work all over the place. For instance, the walls of the city are constantly rearranged. Walls are extended to bulge out or in. Then completely wall-enclosed structures are generated. Think of a little zoo packed with animals and people. These zoos also move around, pulled with the help of giant cranes and tow trucks that use ATP (human work). Eventually the zoo islands reach the city wall. The wall is opened and joined with the zoo wall to release the animals out of the city. There are even hospitals or car workshops if you prefer. Trucks and cars that do not work well are completely disassembled and rebuilt from scratch. The DNA in the library is also constantly rearranged with the help of ATP. Loops are pulled out to copy the library.

We could go on with this analogy for quite a while, but you got the gist. On the one hand, it is a very dynamic city, where things constantly bounce into each other; and if the combination is right, conversions happen. On the other hand, the city is very well organised and compartmentalised into power stations, zoos, the library. Everywhere things fly back and forth, but they start to move in a certain direction when demand removes certain groups. For example, when a dog gets converted into a cat, the opposite can happen as well – a cat gets reconverted into a dog. As a result, the population is stable. However, if a lot of dogs (nutrients) come into the city, more of them will be converted because the chances are higher that a dog gets converted into a cat than vice versa because there are so many new dogs. ATP gets produced only if it is required; otherwise, production slows down. There is not much waste produced or things happening without purpose. However, the cell will never stop altogether. Something is always happening; it is the ultimate city that never sleeps.

There is a lot of information required to maintain the organisation of the city. Although we always relate information in our cells to the genome, there is a lot of additional information in the cell structure. The wall, pipes, zoos, cranes, trucks all represent structural information that helps organising a city. Because information is key to understanding life, we must now investigate information and its role in the generation of life.

We keep our microscopic analogy; let us use sheep this time, representing molecules. We have two paddocks with equal numbers of sheep running around randomly. There is a gate between both paddocks, this time not a kissing gate (Figure 59). Instead, a demon is present who can open or close the gate quickly. The demon observes the sheep and opens the gate when a sheep comes close to the gate in paddock 1 but never opens the gate when a sheep comes close to it in paddock 2. Because the sheep move around randomly, some will pass through the gate when it briefly opens, but only from paddock 1 to 2 because of the demon. Over time, more sheep will accumulate in paddock 2. In physical terms, paddock 2 would have a higher pressure, which is a form of energy.

Figure 59. Analogy to explain energy generation by the Maxwell demon who controls the gate between paddocks 1 and 2. The demon will only open the gate for sheep passing from paddock 1 to 2.

James Clark Maxwell (1831–1879) devised this thought experiment to point out some potential problem in the developing discipline of thermodynamics. Thermodynamics was an exciting discipline at the time because it could explain and improve the steam engine and the conversion of its energy into other energies. The demon could potentially make steam without energy. We discussed the second law of thermodynamics in the first chapter of this book. It states that energy dissipates (gets distributed) and does not spontaneously aggregate in one place. Clearly, there is a problem with Maxwell's demon. The demon was eventually exorcised by Leo Szilard (1898–1964) and Leon Brillouin (1889–1969), who recognised that the demon needs information where the sheep are located; and to acquire this information, energy is required.[161] In fact, the energy equals the energy generated by the accumulation of sheep in paddock 2. In his thought experiment, Maxwell assumed the gate to be without friction, so opening and closing would not cost extra energy. This has always been something that bothered me. Gates are a bit different at the molecular level than in a fence because the gate is so tiny that due to thermal energy it will always wobble, opening and closing spontaneously. Remember all the fast-moving molecules in our city analogy. The reason is temperature. The higher the temperature, the more the molecules rattle, and the faster they are. Recall from the first chapter that splitting ATP generates a local heat equivalent to 3,900 °C. That is a lot of rattling. If the molecular gates constantly rattle, it is as likely that sheep pass from paddock 1 to paddock 2 as in the opposite direction. This is perfectly fine, no demon required, and no gradient established out of nowhere. To push or hold the gate in a particular direction, we need energy; and as we saw throughout this book, all our pumps use ATP to do just that. We also saw that our gates contain information. Like a lock, only a certain key fits into the lock. However, the information is not selective in any direction. Let us assume, we have sheep and cows, but the lock is too small for the cows (Figure 60). If we have an equal number of sheep and cows in both paddocks, nothing will happen because it is equally possible for a sheep to pass from paddock 1 to 2 or vice versa. The information is built into the gate, but it is not a demon that can violate the second law of thermodynamics. What happens if we have sheep in paddock 1 and cows in paddock 2? The sheep will eventually end up having equal numbers on both sides, provided the paddocks are large enough to house both groups of animals. This is what happens in reality, and it is called osmosis. It causes plants to have strength and elasticity when provided with water. The sheep just follow the concentration gradient. Eventually the pressure will balance the energy provided by the uneven distribution of the sheep and cows in the beginning. The pressure in our analogy is just the higher number of animals in paddock 2.

We can now improve our understanding of the Maxwell demon. If the gate is indeed friction-less, it is only the information that makes the energy balance. If the gate is real, you also need energy to push the gate in a particular direction, for instance ATP. Biological gates do not gather information about the sheep running around, but they use ATP to push the gate in one direction when a sheep accidentally walks in from the right side. If the sheep walks in from the other end,

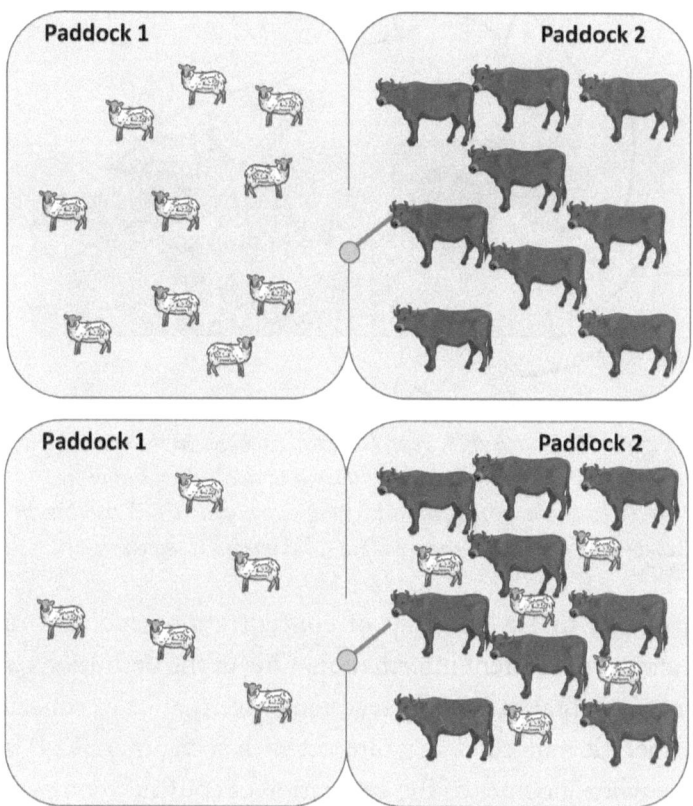

Figure 60. A selective gate for sheep between two paddocks causes the build-up of animal numbers in paddock 2 when starting with even numbers of two different animals.

nothing will happen because the gate is pushed in that direction. To be very precise, there is the occasional slip because of the constant rattling that will allow a sheep to slip through in the opposite direction, but this is an exceedingly rare event. The more energy you put in, the less likely the slip is. We can now understand why macroscopic ratchets, like in Figure 61, do not exist at the molecular level. The catch would be jumping up and down because of thermal movement. As a result, the gear can go both ways. However, with an input of ATP, you could push the catch down, which would favour counterclockwise rotation. As a result, molecular ratches do exist, but they require energy.[1]

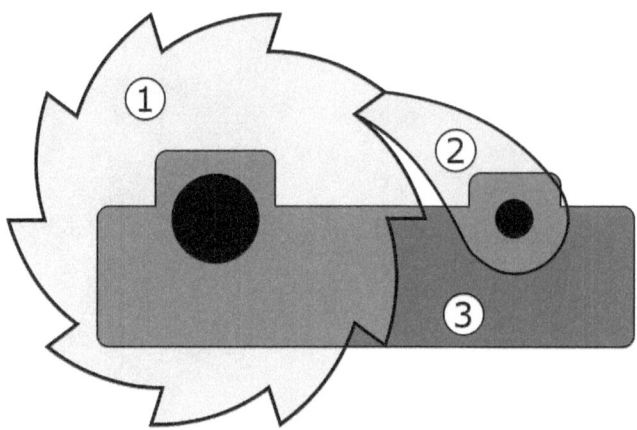

Figure 61. A ratchet: (1) gear, (2) catch, (3) base. In the macroscopic world, this device ensures that only counterclockwise rotation occurs. In the molecular world, the catch jumps up and down all the time if there is no additional energy input (Wikimedia Commons).

We have seen that a build-up of concentration gradients requires energy and gate-inherent information. One of the definitions of life is the compartmentalisation of a reaction space that has a concentrated soup of specific molecules that interact with each other. ATP is at the epicentre when it comes to the generation of confined compartments with concentration gradients as we have seen many times in this book. The gate-inherent information is of course encoded in our DNA. The sequence of a gene encodes the order of amino acids (pearls) in a protein. As the necklace comes out of the protein synthesising nanomachine, called the ribosome, it spontaneously starts to fold into its intricated shape, because of attractive forces between the amino acids and the exclusion of water. By the time the last pearl left the ribosome, the shape of the protein is complete. Some larger proteins or complexes of proteins need some help with the folding, but by and large, it is a spontaneous process.

Let us use our city analogy again but start way back in history to explain the generation of life and the role that information plays. Let us assume that the whole human society represents an organism and that each human is a molecule. Initially, we have groups of hunter-gatherers, and they interact in small ways – exchanging food, going for a hunt, or building a primitive shelter. The shelter could present a primitive protein or a cellular structure. There is limited information available to pass on to the next generation – how to hunt, which food is eatable, which is toxic, how to raise a child. Eventually, animals are domesticated, and some plants are used systematically to provide food. The society turns from hunter-gatherer into an agrarian society. There is much more information to pass on now – when to seed plants, when to harvest, how to harvest, how to preserve food, how to feed animals, how to slaughter. It gets increasingly difficult to pass on all the information, so people become specialised. Eventually, some form of writing is required to pass on information – how much wheat is in storage, when to harvest, how to process? With writing, progress is faster, and settlements can turn into towns. Eventually, written material is stored systematically, first as papyri, later as books. As productivity increases, some people have time to think about items that are not essential to survival. Inventions start to happen. Tools are generated to help harvesting and planting, but also to fight each other. Any invention gets documented, so it is preserved for the future. The Industrial Revolution starts. Steam engines are produced; electricity is exploited as a new type of energy that can be converted into other energies such as light or mechanical energy. Innovative ideas are published in scientific journals, which are distributed between countries. These accelerate the development of ideas and inventions even further. Large cities appear with complex structures such as power plants, sewer systems, schools, hospitals, and public transport. Superstructures, such as governments, are becoming essential. These days we can assemble airplanes that contain 2.3 million parts. Not a single person could assemble a Dreamliner, but a large, organised group of people with information can. Quality control and functional testing is essential to ensure that the final product is working. This requires not only the group of people who assemble the airplane but also all the people who provide the parts and the transport that brings the parts to the factory. Moreover, the people are employed by a company, which needs to pay salaries and taxes, select people for employment, and keep the skills and information to assemble an airplane up to date and to improve it.

There is an obvious similarity between a complex human society and a complex organism. It helps to explain how quite simple conglomerations of molecules (humans) can assemble into complex organisms (towns, societies, countries) if an instruction is available that can be passed on to the next generation. The analogy goes far in some places. For instance, I mentioned that increased productivity gave people more time to think about their environment and to develop innovative ideas. In biological language, a gene can be duplicated by accident, and the second copy can develop into something new without compromising the function of the original gene. Because our DNA preserves the information and passes it on to the next generation, the information can be used to generate something new through mutation, gene duplication, change of gene activity, and so on. One clear difference is the targeted nature of human activity. To assemble the airplane, we look for a certain part and place it where it belongs. In cellular life, molecules bounce around; and when they fit, they are accepted. Even when DNA is copied, all sorts of molecules will bounce into the machinery but do not fit. Only when the right building blocks (the so-called deoxy nucleotides dATP, dCTP, dCTP, and dTTP) come along, one of them fits in place. It is like assembling a puzzle by frantically throwing all pieces into the gap until one fits. The energy of releasing the phosphate attached to the nucleotide is used to join them and to go to the next position of the puzzle and do the same thing all over again. This is the reason it is important that all molecules bounce around so frantically and are concentrated inside the cell, and the concentration process requires ATP, as we saw earlier. It also emphasizes the information content of the structures, because only a fitting part will be able to complete a complex assembly.

This gives us a plausible view of how life became more complex over time once information could be passed on as DNA. But the hard problem is the beginning. How was information laid down in the first place, and even harder how was information retrieved and used? The first microorganisms for which reliable fossils are available were found in the Pilbara region of Western Australia. They have been dated to be 3.5 billion years old. However, we assume that they already stored their information in DNA, used RNA, and made proteins on ribosomes. In our analogy, that is like going back to the ancient Greeks who already wrote down what they knew and had a well-organised state. How do we know that organisms that long ago were already so sophisticated? The idea here is that while many ancient organisms have died out, many continued to flourish and have changed little, because they were already quite adapted to their environment. Bacteria are offspring of more primitive life forms than a dog or a human. Comparing the DNA sequences of many current representatives of life gives us an idea of the genes that are essential for an organism, and these are often relatively similar between distant modern organisms, suggesting that they were also similar a long time ago. One of the simplest organisms, mycoplasma genitalium, has only 482 genes compared to our 21,000 and yet can happily exist in the cosy environment of our genital tracts. Yet these organisms already use massive amounts of ATP. Simple bacteria use 50–55 billion ATP molecules to duplicate from one cell to two cells, which is 50–100 times their own cell mass.

But what is the molecular equivalent of an early agrarian society? As we all know, DNA is a double strand, so the information is preserved well and duplicated easily. However, it is a rigid and stable molecule that cannot be used for much else. Fair enough, we do not want to mess up the library. We copy the DNA information over to RNA, and that is a much more flexible molecule, which can almost fold up like the amino acids in a protein. We depicted this so-called messenger RNA as a human chain, which is similarly flexible. It is intriguing to look at the nanomachine that makes new proteins, called the ribosome (Figure 62). The light-gray spaghetti is RNA, while the black stuff is protein. This is a very ancient protein/RNA composite that can make new proteins provided an additional messenger RNA is available that can be fed into the machine like a punched tape. The ribosome will then stitch together pearls into a necklace in the instructed order. ATP or more broadly the nucleotides ATP, GTP, UTP, and CTP are centre stage in this machine. All RNA is made from those four molecules, losing two phosphates each to join up as A-G-U-C. They bring the energy with them to join up. Although we are getting closer to the beginning of life, it is still a long way to make a ribosome and all the stuff that goes with it. We need amino acids, not as pure amino acids, but in an activated form hanging on to an RNA molecule. We need plenty of proteins to assemble the ribosome and to start stitching the pearls together. We also need a lot of energy. We need ATP to load the amino acids onto the RNA molecule, and we need GTP to let the ribosome march along the mRNA molecule. It is like a three-part egg-and-hen problem. You need an RNA to make a ribosome, you need a ribosome to make a protein, and you need a protein to make the RNA. To simplify our problem, we can go further back and get rid of proteins altogether and rely on curled-up RNA to catalyse reactions including stitching nucleotides together to make RNA. This is possible; RNA enzymes or ribozymes have been known for some time.

In fact, the ribosome is a ribozyme as well.[162,163] As discovered by Thomas Cech and Sidney Altman, the ribozymes are mostly cutting and splicing RNA and DNA, but they can be designed in the lab to facilitate other reactions as well. Most notably RNA molecules have been modified to help cut DNA. This technology, known as CrispR/Cas9, was discovered by Jennifer Doudna and Emmanuelle Charpentier.[164] Thomas Cech and Sidney Altman received the Nobel Prize in 1989, and Jennifer Doudna and Emmanuelle Charpentier received the Nobel Prize in 2020. Ribozymes are not as good as proteins in converting dogs into cats, so they disappeared once better alternatives were available. However, to propagate information, they are quite suitable. If the surface of another RNA molecule could be the template to stitch nucleotides together in the right order, that

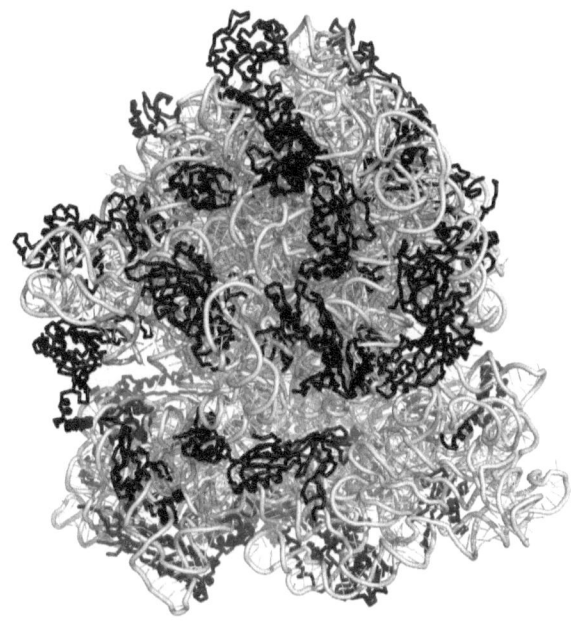

Figure 62. Structure of the ribosome. This nanomachine is made up of RNA (light grey) and protein (black). Both play a critical role in its function.

could become an uncomplicated way of passing information to the next generation. Moreover, a free nucleotide could fit into a gap of an RNA molecule that acts as a catalyst to modify such a molecule. Thus, RNA can be both information and an enzyme. Using evolution in the test tube, it has been possible to select ribozymes that can slowly polymerase RNA along a template.[165]

This scenario has been termed the RNA world.[166] There are still a lot of things unexplained in the RNA world, but it is a plausible earlier scenario of life ascending. How did nucleotides arise out of the clutter of organic molecules?[165] How were nucleotides chained together? How was the RNA copied? How did natural selection give rise to a functional organism?[167] Even if you have a soup of ATP, GTP, UTP, and CTP, nothing much will happen. They do not link together, because it is much more likely just to release a phosphate or two and produce heat.

As a result, the RNA world is already the equivalent of the agrarian society, and we need to go back to the hunter-gatherer equivalent.

This brings us back to metabolism where little information is required to carry out simple reactions. We know that oxygen was generated on earth only after photosynthetic bacteria had evolved. The original atmosphere contained carbon dioxide, methane, hydrogen, nitrogen, and some other gases. Under these conditions, cyanide and formaldehyde could have formed, which are very reactive molecules. Harold Urey (1893–1981) and Stanley Miller (1930–2007) performed the classical experiment in 1952 in which amino acids were synthesised in a primordial atmosphere and soup with the help of electric sparks. It started a whole field of endeavour. Numerous experiments have since been performed to carry out reactions under conditions of the early earth, called the primordial soup. Joan Oro (1923–2004) and A. P. Kimball then succeeded to synthesise adenine. Synthesis of sugars has been accomplished as well, but the tricky part is to put them together to make a nucleotide.[167]

The earth formed 4.6 billion years ago, but for the first 800 million years, it was not conducive to the formation of life. That leaves 'only' about 300 million years for the primordial soup to produce something useful. In Figure 63, I have compared three molecules. One is ATP, the other is Euler's coferment (NADH), and the third is called coenzyme A. We have come across coenzyme A as the entry accelerator to the Arc de Triomphe cycle race. It was discovered by Fritz Lipmann. It is too much of a coincidence that all three molecules that are essential to facilitate the breakdown of nutrients and to generate ATP are built up on the same core, namely ADP (Figure 63). And the same core is also found in RNA and DNA. Clearly there was an original molecule that was partially modified and then able to serve mutliple purposes. What reaction could have generated these molecules? We require energy to make such molecules, but such ecosystems are available in hot springs. Another problem is the availability of phosphate. There is plenty of phosphate on our planet, but it is typically found in minerals that do not dissolve in water. The opposite problem applies to calcium ions. They are abundant but don't work well together with ATP. As ATP is used to provide energy, calcium ions must be excluded to avoid formation of insoluble complexes.[168]

In the absence of oxygen, life resorts to a simpler lifestyle called fermentation. You need far fewer genes to carry out fermentations. The urea cycle that we met when discussing the role of ATP in liver can be redesigned into a linear simple fermentation pathway starting with arginine that could make ATP from ADP. L. H. Stickland in 1934 found a reaction series where one amino acid helps to convert another amino acid. In this reaction, electrons can change hands without oxygen being involved. In a living organism, we can make ATP with this reaction. This can involve simple amino acids such as glycine, which can be generated in copious quantities in a primordial soup without involving any life. In living organisms, NADH, ATP, and coenzyme A are involved to facilitate the reactions. Thus, ATP and its cousins are found right at the beginning of life. However, it has not been possible to generate ATP under conditions mimicking the primordial soup. Chaining several nucleotides together to form a short RNA is also difficult because water favours the splitting of the resulting molecules. Nevertheless, short pieces of RNA have been generated based on slightly different nucleotides than those we are familiar with today.[167]

Deep ocean vents have been considered for quite some time as places where life could have emerged. Particularly 'lost city hydrothermal fields' provide an environment where methane, hydrogen, and thermal energy are plentiful and stable.[169] The chemistry might have worked in slightly different ways, using hydrogen sulphide to facilitate reactions. Günter Wächtershäuser developed several reactions to account for the generation of early biomolecules.[170] He also proposed that metabolism was occurring before the generation of the RNA world. There is a rich bacterial environment found in deep ocean vents today, which use these energy sources. Their metabolism is similar to assumed ancient bacteria.[171] There is plenty of energy around and hydrogen which is required in addition to carbon dioxide to generate organic molecules. A simple pathway of carbon dioxide fixation has been discovered that makes use of the key molecules depicted in Figure 63. It is another cyclic pathway which generates a molecule of two carbons, which is reactive enough to generate further biomolecules.[172] However, incorporation of carbon dioxide into an organic molecule is energetically costly, always requiring ATP because carbon dioxide is such a stable molecule. It is possible though that carbon monoxide, which is much more reactive, could have been used as an alternative.

As an alternative, a reverse Arc de Triomphe cycle has been proposed[170,173] as an early metabolic pathway but also refuted for its unlikely chemistry.[174]

The vents also offer significant pH gradients between the interior of the vents, which is alkaline compared to the pH of sea water, which is slightly acidic. This could form the basis of chemiosmotic ATP production, but it is unclear how the proteins would be generated to carry the reactions out.

Moreover, ribose, which is the sugar found in DNA, RNA, ATP, NADH, and coenzyme A, is quite unstable at higher temperatures. The phosphate bond, which is so important to life, is as we learned also quite unstable.

Another hard problem is the membrane around cells.[175] As we learned, it is essential to generate high local concentrations of reactants, but at the same time it is a barrier. You need kissing gates in the membrane to let nutrients in and out selectively, but for that we already must have proteins. It is not possible to put an RNA molecule across a membrane unless it is coated with specific detergents. Another possible solution is that life emerged in hot springs on land, like

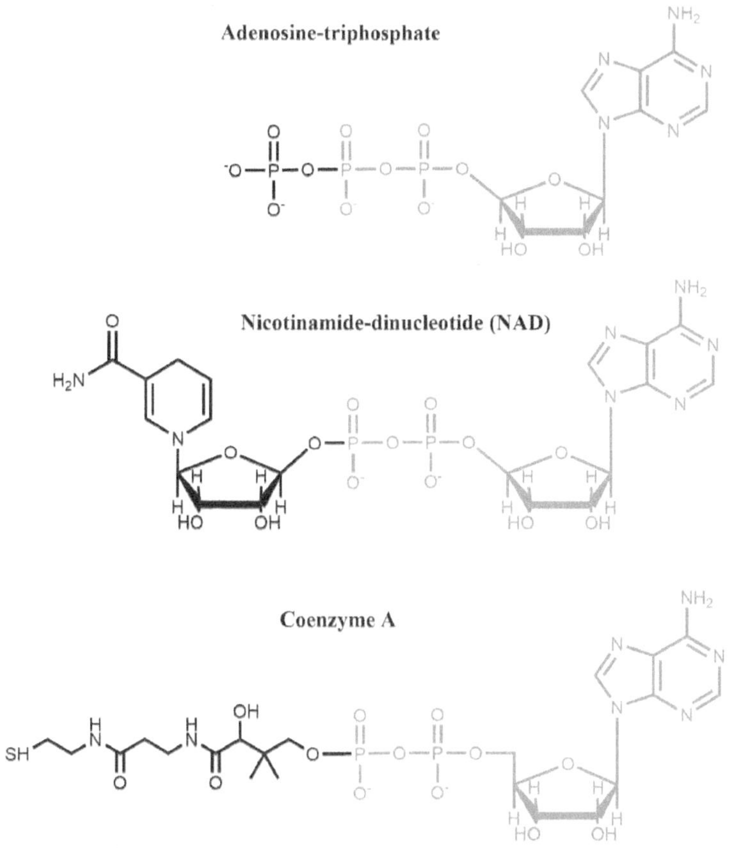

Figure 63. Ancient cofactors that catalyse principal chemical reactions in organisms. Note that the right half of all three molecules is the same.

those in Yellowstone National Park. There is plenty of energy, but also a confined reaction space. Mud can dry out and then rehydrate again. Such cycles can facilitate the generation of polymers. David Deamer succeeded in polymerising nucleotides in plausible primordial conditions that involve lipids.[176] The conditions are sort of plausible, because in the laboratory scientists must compress processes that took thousands of years into days. Importantly, instead of using ATP (or its cousins GTP, CTP, and UTP), he used AMP, which does not have high-energy phosphates.

This, at least partially, solves one of the biggest conundrums for any theory on how life could have evolved. The high-energy bond in ATP is prone to splitting because it has a high energy content, and the products are much more stable. The presence of water facilitates the splitting. However, AMP or related nucleotides can be polymerised in an environment where water activity is reduced, such as a pond drying out and then filling with water again. Periodically changing reaction conditions are probably essential for the generation of life, because they can provide the energy required to overcome the tendency to increase random distribution. We also saw that we need information to generate life. This could be initially provided by structured surfaces that could facilitate certain reactions. We saw that a lock and key can provide enough information to build a concentration gradient if energy is provided.

The simple fermentation reactions are probably the equivalent to the hunter-gatherer society, but the jump to the involvement of an ATP- or AMP-like molecule is a big one, and the next step to stitch these together might involve cycles of dry and wet. Then we need a cell membrane around it. Lastly and most importantly, we need the Maxwell demon (a.k.a. information) to build structures that can be given to the next generation of primitive life.

Do not get me wrong. I am not religious, but it is a hard problem that remains largely unresolved. Nevertheless, researchers are coming up with possible schemes, step by step. What we know is that it works today and that it took 300 million years to develop. Even when we speed up reactions in the laboratory, that is a long time. However, ATP (or AMP) was there right at the very beginning. It might have been preceded by chemically similar coferments involved in oxidation and reduction reactions,[171] but to polymerise nucleotides into RNA, it must have been an established molecule.

If life has evolved somewhere else in the universe, for which we do not have yet evidence, would it also have ATP? Not necessarily, because other molecules have energy-rich bonds and could replace ATP. However, organic chemistry has a lot of hard and fast rules on how molecules can react with each other, and it appears likely that similar molecules such as amino acids and organic acids would be used in extra-terrestrial life. If we ever detect another blue planet in the universe, it is likely that life evolved more than once. This does not matter for us. We still must figure out how life evolved on this planet, and this required ATP right at the very beginning.

12

Epilogue

Australians like to say, 'There is no free lunch,' and I wholeheartedly agree. By the time lunch is served, the cow used a lot of ATP to provide the beef, the plants used ATP to provide the veggies, and a lot of people in between used ATP to prepare the ingredients and the meal. If anyone ever asks you, 'What is life?' you can answer, 'If it makes ATP, it is life.'

I hope I have made the case for ATP. It does deserve its own book, and even non-experts will appreciate its crucial role for life on earth as we know it. Although I have been generous, I was surprised how many Nobel Prizes were touched by ATP or directly involve ATP. The announcement of Nobel laureates each year in October builds a bridge between science and the public for a brief moment. Media rush in to explain what has been done, but the news fades quickly. By connecting the dots of so many Nobel Prizes, we can appreciate how science develops. Many of the prizes in the first half of the twentieth century were awarded for fundamental insights without commercial interest. In the second half of the twentieth century, this changed dramatically. We looked at medical imaging and blockbuster drugs. Many important discoveries were not honoured with Nobel Prizes and yet have changed the lives of many people. The analogy of the development of human society as a complex organism shows how important the inheritance of information is. This information allows us to develop new medicines and technology to improve the lives on our planet. This allows us to reduce the entropy locally to assemble complex structures, but it comes at the expense of increasing entropy in our environment. Figure 64 shows an image of me next to Ludwig Boltzmann in the courtyard of the University of Vienna. Ludwig Boltzmann established the relationship between order and entropy. As we saw, energy is required to generate order and information and to store it. On our planet, ATP is essential and centre stage to do just that.

Figure 64: The author next to the sculpture of Ludwig Boltzmann in the courtyard of the University of Vienna.

References

1   Hoffmann, P. M. *Life's Ratchet*. (Basic Books, 2012).
2   Schatz, G. *Feuersucher*. (Wiley-VCH, 2011).
3   Brown, G. *The energy of life*. (Harper Collins Publishers, 1999).
4   McElroy, W. D. The Energy Source for Bioluminescence in an Isolated System. *Proc Natl Acad Sci U S A* **33**, 342-345, doi:10.1073/pnas.33.11.342 (1947).
5   Ball, E. G. in *A symposium on respiratory enzymes, Madison, The University of Wisconsin Press*.
6   Bell, M. S. *Lavoisier in the Year One: the Birth of a New Science in an Age of Revolution*. (WW Norton & Company, 2005).
7   Culotta, C. A. Tissue oxidation and theoretical physiology: Bernard Ludwig, and Pfluger. *Bulletin of the History of Medicine* **44**, 109-140 (1970).
8   Holmes, F. L. *Between biology and medicine: The formation of intermediary metabolism*. Vol. 12 (University of California Office for History of Science and Technology, 1992).
9   Liebig, J. Ueber die Bestandtheile der Flüssigkeiten des Fleisches. *Justus Liebigs Annalen der Chemie* **62**, 257-369, doi:https://doi.org/10.1002/jlac.18470620302 (1847).
10 Holmes, F. L. Elementary Analysis and the Origins of Physiological Chemistry. *Isis* **54**, 50-81, doi:Doi 10.1086/349664 (1963).
11 Holmes, F. L. Claude Bernard, the milieu interieur, and regulatory physiology. *Hist Philos Life Sci* **8**, 3-25 (1986).
12 Lagerkvist, U. *Enigma Of Ferment, The: From The Philosopher's Stone To The First Biochemical Nobel Prize*. (World Scientific, 2005).
13 Buchner, E. Alkoholische Gährung ohne Hefezellen. *Berichte der deutschen chemischen Gesellschaft* **30**, 117-124, doi:https://doi.org/10.1002/cber.18970300121 (1897).
14 Harden, A. & Young, W. J. The alcoholic fermentation of yeast-juice. *Proc. R. Soc. Lond. B* **77**, 405-420, doi:https://doi.org/10.1098/rspb.1906.0029 (1906).
15 Harden, A. & Young, W. J. The alcoholic ferment of yeast-juice. Part III.-The function of phosphates in the fermentation of glucose by yeast-juice. *Proc. R. Soc. Lond. B.* **80**, 299-311, doi:https://doi.org/10.1098/rspb.1908.0029 (1908).
16 NobelPrize.org. *Arthur Harden – Biographical*, <<https://www.nobelprize.org/prizes/chemistry/1929/harden/biographical/>> (2022).

17   Nachmansohn, D. Biochemistry as part of my life. *Annu Rev Biochem* **41**, 1-28, doi:10.1146/annurev.bi.41.070172.000245 (1972).

18   NobelPrize.org. *Otto Warburg – Biographical*, <<https://www.nobelprize.org/prizes/medicine/1931/warburg/biographical/>> (2022).

19   Peters, R. A. Otto Meyerhof, 1884 - 1951. *Obit. Not. Fell. R. Soc.9174–200*, doi:http://doi.org/10.1098/rsbm.1954.0013 (1954).

20   States, D. M. Otto Meyerhof and the Physiology Institute: the Birth of Modern Biochemistry. *NobelPrize.org* (2021).

21   Emmerich, M. Cells flexing their muscles. *Max Planck Research* **1**, 86-87 (2009).

22   Parnas, J. Obituary Prof. G. Embden. *Nature*, 994-995 (1933).

23   Fletcher, W. M. & Hopkins, F. G. Lactic acid in amphibian muscle. *J Physiol* **35**, 247-309, doi:10.1113/jphysiol.1907.sp001194 (1907).

24   Young, F. G. Claude Bernard and the discovery of glycogen; a century of retrospect. *Br Med J* **1**, 1431-1437, doi:10.1136/bmj.1.5033.1431 (1957).

25   Embden, G. & Zimmermann, M. Über die Chemie des Lactacidogens. *Biol. Chem.* **167**, 114 (1927).

26   Parnas, J., Oostern, P. & Mann, T. Linkage of Chemical Changes in Muscle. Nature 134, 1007 (1934). https://doi.org/10.1038/1341007a0. *Nature*, 1007 (1934).

27   Langen, P. & Hucho, F. Karl Lohmann and the discovery of ATP. *Angew Chem Int Ed Engl* **47**, 1824-1827, doi:10.1002/anie.200702929 (2008).

28   Eggleton, P. & Eggleton, G. P. The Inorganic Phosphate and a Labile Form of Organic Phosphate in the Gastrocnemius of the Frog. *Biochem J* **21**, 190-195, doi:10.1042/bj0210190 (1927).

29   Fiske, C. H. & Subbarow, Y. The Nature of the "Inorganic Phosphate" in Voluntary Muscle. *Science* **65**, 401-403, doi:10.1126/science.65.1686.401 (1927).

30   Hill, A. V. The revolution in muscle physiology. *Physiological Reviews* **12**, 56-67 (1932).

31   Embden, G., Hirsch-Kauffmann, H., Lehnartz, E. & Deuticke, H. J. Über den Verlauf der Milchsäurebildung beim Tetanus. *Biological Chemistry* **151**, 209-231, doi:https://doi.org/10.1515/bchm2.1926.151.4-6.209 (1926).

32   Lohmann, K. Über die pyrophosphatfraktion im muskel. *Naturwissenschaften* **17**, 624-625 (1929).

33   Fiske, C. H. & Subbarow, Y. Phosphorus compounds of muscle and liver. *Science* **70**, 381-382 (1929).

34   Mukheerjee, S. *The Emperor of all Maladies*. 30-31 (Scribner, 2010).

35   Meyerhof, O., Lohmann, K. & Meyer, K. Über das Koferment der Milchsäurebildung im Muskel. *Biochem. Z.* **237**, 437-444 (1931).

36   Embden, G., Deuticke, H. J. & Kraft, G. Über die intermediären Vorgänge

bei der Glykolyse in der Muskulatur. *Klinische Wochenschrift* **12**, 213-215 (1933).

37 Meyerhof, O. 337 (Nature Publishing Group, 1933).

38 Maruyama, K. The Discovery of Adenosine-Triphosphate and the Establishment of Its Structure. *J Hist Biol* **24**, 145-154 (1991).

39 Negelein, E. & Brömel, H. *Biochem Z.* **301**, 135 (1939).

40 Lohmann, K. Über die enzymatische Aufspaltung der Kreatinphosphorsäure; zugleich ein Beitrag zum Chemismus der Muskelkontraktion. *Biochem. z* **271**, 264-277 (1934).

41 Lipmann, F. Metabolic generation and utilization of phosphate bond energy. *Advances in enzymology and related areas of molecular biology* **1**, 99-162 (1941).

42 Lipmann, F. A long life in times of great upheaval. *Annu Rev Biochem* **53**, 1-33, doi:10.1146/annurev.bi.53.070184.000245 (1984).

43 Florkin, M. & Stotz, E. H. *A History of Biochemistry Part III. History of the Identification of the sources of free energy in organisms*. Vol. 31 (Elsevier Scientific Publishing Company, 1975).

44 Kalckar, H. M. The nature of energetic coupling in biological synthesis. *Chem. Rev.* **28**, 71-178 (1941).

45 Slater, E. C. Keilin, cytochrome, and the respiratory chain. *J Biol Chem* **278**, 16455-16461, doi:10.1074/jbc.X200011200 (2003).

46 Krebs, H. A. Otto Heinrich Warburg, 1883-1970 *Biogr. Mems Fell. R. Soc.* **18**, 18628–18699 (1972).

47 Cori, C. F. & Cori, G. T. Glycogen Formation in the liver from D- and L-lactic acid. *Journal of Biological Chemistry* **81**, 389-403, doi:10.1016/S0021-9258(18)83822-4 (1929).

48 Lassen, J. *What does it feel like to run a marathon?*, <https://www.quora.com/What-does-it-feel-like-to-run-a-marathon> (2018).

49 Wigglesworth, V. B. The utilization of reserve substances in Drosophila during flight. *J Exp Biol* **26**, 150-163, illust (1949).

50 Lane, N. *Life Ascending*. 88-117 (Profile Books Ltd, 2010).

51 Claude, A. in *Harvey Society Lectures 44* (The Rockefeller University, 1948).

52 Krebs, H. A. The history of the tricarboxylic acid cycle. *Perspectives in Biology and Medicine* **14**, 154-172 (1970).

53 Manchester, K. L. Albert Szent-Gyorgyi and the unravelling of biological oxidation. *Trends Biochem Sci* **23**, 37-40, doi:10.1016/s0968-0004(97)01167-5 (1998).

54 Obatomi, D. K. & Bach, P. H. Biochemistry and toxicology of the diterpenoid glycoside atractyloside. *Food Chem Toxicol* **36**, 335-346, doi:10.1016/s0278-6915(98)00002-7 (1998).

55 Anwar, M., Kasper, A., Steck, A. R. & Schier, J. G. Bongkrekic Acid-a Review of a Lesser-Known Mitochondrial Toxin. *J Med Toxicol* **13**, 173-179,

doi:10.1007/s13181-016-0577-1 (2017).

56 Klingenberg, M. When a common problem meets an ingenious mind. *EMBO Rep* **6**, 797-800, doi:10.1038/sj.embor.7400520 (2005).

57 Lehninger, A. L. Phosphorylation coupled to oxidation of dihydrodiphosphopyridine nucleotide. *J Biol Chem* **190**, 345-359 (1951).

58 De Meis, L. How Enzymes Handle the Energy Derived from the Cleavage of High-energy Phosphate Compounds. *J Biol Chem* **287**, 16987-17005, doi:https://doi.org/10.1074/jbc.X112.363200 (2012).

59 Prebble, J. Peter Mitchell and the ox phos wars. *Trends Biochem Sci* **27**, 209-212, doi:10.1016/s0968-0004(02)02059-5 (2002).

60 Boyer, P. D. *et al.* Oxidative phosphorylation and photophosphorylation. *Annu Rev Biochem* **46**, 955-966, doi:10.1146/annurev.bi.46.070177.004515 (1977).

61 Engelhardt, V. A. & Lyubimova, M. N. Myosin and adenosinetriphosphatase. *Nature* **144**, 668-669 (1939).

62 Szent-Gyorgyi, A. G. The early history of the biochemistry of muscle contraction. *J Gen Physiol* **123**, 631-641, doi:10.1085/jgp.200409091 (2004).

63 Huxley, H. E. Fifty years of muscle and the sliding filament hypothesis. *Eur J Biochem* **271**, 1403-1415, doi:10.1111/j.1432-1033.2004.04044.x (2004).

64 Hasselbach, W. & Makinose, M. [The calcium pump of the "relaxing granules" of muscle and its dependence on ATP-splitting]. *Biochem Z* **333**, 518-528 (1961).

65 Valenstein, E. S. *The war of the soups and the sparks.* 51-67 (Columbia University Press, 2005).

66 Dale, H. H. & Gaddum, J. H. Reactions of denervated voluntary muscle, and their bearing on the mode of action of parasympathetic and related nerves. *J Physiol* **70**, 109-144, doi:10.1113/jphysiol.1930.sp002682 (1930).

67 MacLaren, D. & Morton, J. *Biochemistry for Sport and Exercise Metabolism.* 2-10 (John Wiley & Sons Ltd, 2012).

68 Cheung, K., Hume, P. & Maxwell, L. Delayed onset muscle soreness: treatment strategies and performance factors. *Sports Med* **33**, 145-164, doi:10.2165/00007256-200333020-00005 (2003).

69 Ballard, H. J. ATP and adenosine in the regulation of skeletal muscle blood flow during exercise. *Sheng Li Xue Bao* **66**, 67-78 (2014).

70 Hardie, D. G. & Sakamoto, K. AMPK: a key sensor of fuel and energy status in skeletal muscle. *Physiology (Bethesda)* **21**, 48-60, doi:10.1152/physiol.00044.2005 (2006).

71 Aird, W. C. Discovery of the cardiovascular system: from Galen to William Harvey. *J Thromb Haemost* **9 Suppl 1**, 118-129, doi:10.1111/j.1538-7836.2011.04312.x (2011).

72 Pinnell, J., Turner, S. & Howell, S. Cardiac muscle physiology. *Continuing*

Education in Anesthesia, Critical Care & Pain **7**, 85-88 (2007).

73 Stanley, W. C., Recchia, F. A. & Lopaschuk, G. D. Myocardial substrate metabolism in the normal and failing heart. *Physiol Rev* **85**, 1093-1129, doi:10.1152/physrev.00006.2004 (2005).

74 Dunn, J.-O. C., Mythen, M. G. & Grocott, M. P. Physiology of oxygen transport. *BJA Education* **16**, 341-348, doi:10.1093/bjaed/mkw012 (2016).

75 Moore, L. G. Measuring high-altitude adaptation. *J Appl Physiol (1985)* **123**, 1371-1385, doi:10.1152/japplphysiol.00321.2017 (2017).

76 Ashcroft, F. M. *Life at the extremes.* (Harper Collins, 2000).

77 Davis, R. W. A review of the multi-level adaptations for maximizing aerobic dive duration in marine mammals: from biochemistry to behavior. *J Comp Physiol B* **184**, 23-53, doi:10.1007/s00360-013-0782-z (2014).

78 Ferry, G. *Max Perutz and the secret of life.* (2007).

79 Arthurs, G. J. Carbon dioxide transport. *Continuing Education in Anesthesia, Critical Care & Pain* **5**, 207-210 (2005).

80 Taha, M. & Lopaschuk, G. D. Alterations in energy metabolism in cardiomyopathies. *Ann Med* **39**, 594-607, doi:10.1080/07853890701618305 (2007).

81 Kalogeris, T., Baines, C. P., Krenz, M. & Korthuis, R. J. Cell biology of ischemia/reperfusion injury. *Int Rev Cell Mol Biol* **298**, 229-317, doi:10.1016/B978-0-12-394309-5.00006-7 (2012).

82 Bear, M. F., Connors, B. W. & Paradiso, M. A. *Neuroscience: Exploring the brain.* (Williams & Wilkins, 1996).

83 Millett, D. Hans Berger: from psychic energy to the EEG. *Perspect Biol Med* **44**, 522-542, doi:10.1353/pbm.2001.0070 (2001).

84 Cowan, W. M. & Kandel, E. R. in *Synapses* (eds W.M. Cowan, T. Sudhof, & C.F. Stevens) Ch. 1, 1-87 (The Johns Hopkins University Press, 2001).

85 Hodgkin, A. L. & Keynes, R. D. Active transport of cations in giant axons from Sepia and Loligo. *J Physiol* **128**, 28-60, doi:10.1113/jphysiol.1955.sp005290 (1955).

86 Caldwell, P. C., Hodgkin, A. L., Keynes, R. D. & Shaw, T. L. The effects of injecting 'energy-rich' phosphate compounds on the active transport of ions in the giant axons of Loligo. *J Physiol* **152**, 561-590, doi:10.1113/jphysiol.1960.sp006509 (1960).

87 Attwell, D. & Laughlin, S. B. An energy budget for signaling in the grey matter of the brain. *J Cereb Blood Flow Metab* **21**, 1133-1145, doi:10.1097/00004647-200110000-00001 (2001).

88 Watkins, J. C. & Jane, D. E. The glutamate story. *Br J Pharmacol* **147 Suppl 1**, S100-108, doi:10.1038/sj.bjp.0706444 (2006).

89 Khakh, B. S. & Burnstock, G. The double life of ATP. *Sci Am* **301**, 84-90, 92, doi:10.1038/scientificamerican1209-84 (2009).

90 Snyder, S. H. & Pasternak, G. W. Historical review: Opioid receptors. *Trends Pharmacol Sci* **24**, 198-205, doi:10.1016/S0165-6147(03)00066-X (2003).

91 Newman, A. J. Functional magnetic resonance imaging (fMRI). *Research Methods in Second Language Psycholinguistics. Edited by Jill Jegerski and Bill VanPatten*, 153-184 (2013).

92 Raichle, M. E. & Mintun, M. A. Brain work and brain imaging. *Annu Rev Neurosci* **29**, 449-476, doi:10.1146/annurev.neuro.29.051605.112819 (2006).

93 Meyers, M. A. *Prize Fight: the Race and the Rivalry to be the First in Science.*, (Palgrave MacMillan, 2012).

94 Kandel, E. R. *Kandel, Eric R. In search of memory: The emergence of a new science of mind.* . (WW Norton & Company, 2007).

95 Aston-Jones, G. & Cohen, J. D. An integrative theory of locus coeruleus-norepinephrine function: adaptive gain and optimal performance. *Annu Rev Neurosci* **28**, 403-450, doi:10.1146/annurev.neuro.28.061604.135709 (2005).

96 Iversen, L. Julius Axelrod 30 May 1912 — 29 December 2004. *Biogr. Mems Fell. R. Soc.*, 521–531, doi:http://doi.org/10.1098/rsbm.2006.0002 (2006).

97 Feldberg, W. & Sherwood, S. L. Injections of drugs into the lateral ventricle of the cat. *J Physiol* **123**, 148-167, doi:10.1113/jphysiol.1954.sp005040 (1954).

98 Huang, Z. L., Zhang, Z. & Qu, W. M. Roles of adenosine and its receptors in sleep-wake regulation. *Int Rev Neurobiol* **119**, 349-371, doi:10.1016/B978-0-12-801022-8.00014-3 (2014).

99 Retey, J. V. *et al.* A functional genetic variation of adenosine deaminase affects the duration and intensity of deep sleep in humans. *Proc Natl Acad Sci U S A* **102**, 15676-15681, doi:10.1073/pnas.0505414102 (2005).

100 Sims, N. R. & Muyderman, H. Mitochondria, oxidative metabolism and cell death in stroke. *Biochim Biophys Acta* **1802**, 80-91, doi:10.1016/j.bbadis.2009.09.003 (2010).

101 Cox, D. W., Morris, P. G., Feeney, J. & Bachelard, H. S. 31P-n.m.r. studies on cerebral energy metabolism under conditions of hypoglycaemia and hypoxia in vitro. *Biochem J* **212**, 365-370, doi:10.1042/bj2120365 (1983).

102 Stein, Z., Susser, M., Saenger, G. & Marolla, F. *Famine and human development: The Dutch hunger winter of 1944-1945.* (Oxford University Press, 1975).

103 Klein, S., Gastaldelli, A., Yki-Jarvinen, H. & Scherer, P. E. Why does obesity cause diabetes? *Cell Metab* **34**, 11-20, doi:10.1016/j.cmet.2021.12.012 (2022).

104 Tucker, T. *The great starvation experiment: Ancel Keys and the men who starved for science.* (U of Minnesota Press, 2007).

105 Kalm, L. M. & Semba, R. D. They starved so that others be better fed: remembering Ancel Keys and the Minnesota experiment. *J Nutr* **135**, 1347-1352, doi:10.1093/jn/135.6.1347 (2005).

106 Muller, M. J. *et al.* Metabolic adaptation to caloric restriction and subsequent refeeding: the Minnesota Starvation Experiment revisited. *Am J Clin Nutr* **102**, 807-819, doi:10.3945/ajcn.115.109173 (2015).

107 Cahill, G. F., Jr. Starvation in man. *N Engl J Med* **282**, 668-675, doi:10.1056/NEJM197003192821209 (1970).

108 Cahill, G. F., Jr. Fuel metabolism in starvation. *Annu Rev Nutr* **26**, 1-22, doi:10.1146/annurev.nutr.26.061505.111258 (2006).

109 Enerback, S. Human brown adipose tissue. *Cell Metab* **11**, 248-252, doi:10.1016/j.cmet.2010.03.008 (2010).

110 Cannon, B. & Nedergaard, J. The biochemistry of an inefficient tissue: brown adipose tissue. *Essays Biochem* **20**, 110-164 (1985).

111 Bostrom, P. *et al.* A PGC1-alpha-dependent myokine that drives brown-fat-like development of white fat and thermogenesis. *Nature* **481**, 463-468, doi:10.1038/nature10777 (2012).

112 Timmons, J. A., Baar, K., Davidsen, P. K. & Atherton, P. J. Is irisin a human exercise gene? *Nature* **488**, E9-10; discussion E10-11, doi:10.1038/nature11364 (2012).

113 Albrecht, E. *et al.* Irisin - a myth rather than an exercise-inducible myokine. *Sci Rep* **5**, 8889, doi:10.1038/srep08889 (2015).

114 Grundlingh, J., Dargan, P. I., El-Zanfaly, M. & Wood, D. M. 2,4-dinitrophenol (DNP): a weight loss agent with significant acute toxicity and risk of death. *J Med Toxicol* **7**, 205-212, doi:10.1007/s13181-011-0162-6 (2011).

115 Axelrod, C. L. *et al.* BAM15-mediated mitochondrial uncoupling protects against obesity and improves glycemic control. *EMBO Mol Med* **12**, e12088, doi:10.15252/emmm.202012088 (2020).

116 Fruton, J. S. A history of pepsin and related enzymes. *Q Rev Biol* **77**, 127-147, doi:10.1086/340729 (2002).

117 Li, J. J. *Blockbuster drugs: The rise and fall of the pharmaceutical industry.* . (Oxford University Press, 2014).

118 Matthews, D. M. *Protein absorption: development and present state of the subject.*, (Wiley-Liss Inc., 1990).

119 Hamilton, K. L. Robert K. Crane-Na(+)-glucose cotransporter to cure? *Front Physiol* **4**, 53, doi:10.3389/fphys.2013.00053 (2013).

120 Crane, R. K. Robert Kellogg Crane: a scientist remembers. *IUBMB Life* **62**, 642-645, doi:10.1002/iub.366 (2010).

121 Kinter, W. B. & Wilson, T. H. Autoradiographic Study of Sugar and Amino Acid Absorption by Everted Sacs of Hamster Intestine. *J Cell Biol* **25**, 19-39, doi:10.1083/jcb.25.2.19 (1965).

122 Dean, R. B. in *Biol. Symp* Vol. 3 331-348 (1941).

123 Larsen, H. I. Hans Henriksen Ussing. 30 December 1911 — 22 December

2000. *Biogr. Mems Fell. R. Soc.* **55**, 305–335, doi:http://doi.org/10.1098/rsbm.2009.0002 (2009).

124 Kunze, W. A. & Furness, J. B. The enteric nervous system and regulation of intestinal motility. *Annu Rev Physiol* **61**, 117-142, doi:10.1146/annurev.physiol.61.1.117 (1999).

125 Walker, A. M. & Hudson, C. L. The reabsorption of glucose from the renal tubule in amphibia and the action of phlorhizin upon it. *American Journal of Physiology-Legacy Content* **118**, 130-143 (1936).

126 Kleinzeller, A., Kolinska, J. & Benes, I. Transport of glucose and galactose in kidney-cortex cells. *Biochem J* **104**, 843-851, doi:10.1042/bj1040843 (1967).

127 Sutherland, E. W. Nobel Lecture. *NobelPrize.org.* **Nobel Prize Outreach** (2022).

128 Utter, M. F. Pathways of phosphoenolpyruvate synthesis in glycogenesis. *Iowa State J. Sci* **38**, 97-113 (1963).

129 Wood, H. G. & Hanson, R. W. Merton Franklin Utter: March 23, 1917-November 28, 1980. *Biographical memoirs. National Academy of Sciences (US)* **56**, 475-499 (1986).

130 Lewis, G. H. *Obituary Salih J. Wakil PhD*, <https://www.dignitymemorial.com/en-ca/obituaries/houston-tx/salih-wakil-8777028> (2019).

131 Krebs, H. A. & Decker, K. Feodor Lynen, 6 April 1911 - 6 August 1979. *Biogr. Mems Fell. R. Soc.* **28**, 261-317, doi:doi.org/10.1098/rsbm.1982.0012 (1982).

132 Nickelsen, K. & Graßhoff, G. in *Going amiss in experimental research* 91-117 (Springer, 2009).

133 Szakacs, G., Varadi, A., Ozvegy-Laczka, C. & Sarkadi, B. The role of ABC transporters in drug absorption, distribution, metabolism, excretion and toxicity (ADME-Tox). *Drug Discov Today* **13**, 379-393, doi:10.1016/j.drudis.2007.12.010 (2008).

134 Jonker, J. W. *et al.* The breast cancer resistance protein protects against a major chlorophyll-derived dietary phototoxin and protoporphyria. *Proc Natl Acad Sci U S A* **99**, 15649-15654, doi:10.1073/pnas.202607599 (2002).

135 Dietrich, C. G., Geier, A. & Oude Elferink, R. P. ABC of oral bioavailability: transporters as gatekeepers in the gut. *Gut* **52**, 1788-1795, doi:10.1136/gut.52.12.1788 (2003).

136 Roulet, A. *et al.* MDR1-deficient genotype in Collie dogs hypersensitive to the P-glycoprotein substrate ivermectin. *Eur J Pharmacol* **460**, 85-91, doi:10.1016/s0014-2999(02)02955-2 (2003).

137 Ahmed, A. M. History of diabetes mellitus. *Saudi Med J* **23**, 373-378 (2002).

138 Vecchio, I., Tornali, C., Bragazzi, N. L. & Martini, M. The Discovery of Insulin: An Important Milestone in the History of Medicine. *Front Endocrinol (Lausanne)* **9**, 613, doi:10.3389/fendo.2018.00613 (2018).

139 Cooper, T. & Ainsberg, A. *Breakthrough: Elizabeth Hughes, the discovery of insulin, and the making of a medical miracle.* (St. Martin's Press, 2010).

140 Houssay, B. A. Diabetes as a disturbance of endocrine regulation. *Am. J. Med. Sci.* **193**, 581-606 (1937).

141 Dean, P. M. & Matthews, E. K. Electrical activity in pancreatic islet cells. *Nature* **219**, 389-390, doi:10.1038/219389a0 (1968).

142 Henquin, J. C. D-glucose inhibits potassium efflux from pancreatic islet cells. *Nature* **271**, 271-273, doi:10.1038/271271a0 (1978).

143 Rorsman, P. & Trube, G. Glucose dependent K+-channels in pancreatic beta-cells are regulated by intracellular ATP. *Pflugers Arch* **405**, 305-309, doi:10.1007/BF00595682 (1985).

144 Malaisse, W. J., Sener, A., Herchuelz, A. & Hutton, J. C. Insulin release: the fuel hypothesis. *Metabolism* **28**, 373-386, doi:10.1016/0026-0495(79)90111-2 (1979).

145 Ashcroft, F. M. The Walter B. Cannon Physiology in Perspective Lecture, 2007. ATP-sensitive K+ channels and disease: from molecule to malady. *Am J Physiol Endocrinol Metab* **293**, E880-889, doi:10.1152/ajpendo.00348.2007 (2007).

146 Hager, T. *The demon under the microscope: from battlefield hospitals to Nazi labs, one doctor's heroic search for the world's first miracle drug.* (Broadway Books, 2006).

147 Henquin, J. C. The fiftieth anniversary of hypoglycaemic sulphonamides. How did the mother compound work? *Diabetologia* **35**, 907-912, doi:10.1007/BF00401417 (1992).

148 Howard, J. A. Dorothy Hodgkin and her contributions to biochemistry. *Nat Rev Mol Cell Biol* **4**, 891-896, doi:10.1038/nrm1243 (2003).

149 Hall, S. S. Invisible frontiers: The race to synthesize a human gene. (1987).

150 Hughes, S. S. *Genentech: the beginnings of biotech.* (University of Chicago Press, 2011).

151 Rasmussen, N. *Gene jockeys: Life science and the rise of biotech enterprise.* (JHU Press, 2014).

152 Warburg, O. H. Uber den Stoffwechsel der Tumoren. (1926).

153 Rasko, J. & Power, C. *Flesh made new: The unnatural history and broken promise of stem cells.* (ABC Books, 2021).

154 Stockwell, B. R. *Quest for the cure: The science and stories behind the next generation of medicines.*, (Columbia University Press, 2011).

155 Wapner, J. *The Philadelphia chromosome: a genetic mystery, a lethal cancer, and the improbable invention of a lifesaving treatment.* (The Experiment, 2014).

156 Hyun, S. & Shin, D. Small-Molecule Inhibitors and Degraders Targeting KRAS-Driven Cancers. *Int J Mol Sci* **22**, doi:10.3390/ijms222212142 (2021).

157 Fletcher, J. I., Williams, R. T., Henderson, M. J., Norris, M. D. & Haber, M.

ABC transporters as mediators of drug resistance and contributors to cancer cell biology. *Drug Resist Updat* **26**, 1-9, doi:10.1016/j.drup.2016.03.001 (2016).

158 Gottesman, M. M. & Ling, V. The molecular basis of multidrug resistance in cancer: the early years of P-glycoprotein research. *FEBS Lett* **580**, 998-1009, doi:10.1016/j.febslet.2005.12.060 (2006).

159 Thomas, H. & Coley, H. M. Overcoming multidrug resistance in cancer: an update on the clinical strategy of inhibiting p-glycoprotein. *Cancer Control* **10**, 159-165, doi:10.1177/107327480301000207 (2003).

160 Grote, M. *Membranes to Molecular Machines: Active Matter and the Remaking of Life.* (The University of Chicago Press, 2019).

161 Harold, F. M. *The vital force: A study of bioenergetics.* (W.H. Freeman and Company, 1986).

162 Cech, T. R. Structural biology. The ribosome is a ribozyme. *Science* **289**, 878-879, doi:10.1126/science.289.5481.878 (2000).

163 Steitz, T. A. & Moore, P. B. RNA, the first macromolecular catalyst: the ribosome is a ribozyme. *Trends Biochem Sci* **28**, 411-418, doi:10.1016/S0968-0004(03)00169-5 (2003).

164 Isaacson, W. *The code breaker: Jennifer Doudna, gene editing, and the future of the human race.* (Simon and Schuster, 2021).

165 Joyce, G. F. The antiquity of RNA-based evolution. *Nature* **418**, 214-221, doi:10.1038/418214a (2002).

166 Gilbert, W. Origin of life: The RNA world. *nature* **319**, 618-618 (1986).

167 Orgel, L. E. Prebiotic chemistry and the origin of the RNA world. *Crit Rev Biochem Mol Biol* **39**, 99-123, doi:10.1080/10409230490460765 (2004).

168 Plattner, H. & Verkhratsky, A. Inseparable tandem: evolution chooses ATP and $Ca2+$ to control life, death and cellular signalling. *Philos Trans R Soc Lond B Biol Sci* **371**, doi:10.1098/rstb.2015.0419 (2016).

169 Martin, W., Baross, J., Kelley, D. & Russell, M. J. Hydrothermal vents and the origin of life. *Nat Rev Microbiol* **6**, 805-814, doi:10.1038/nrmicro1991 (2008).

170 Wachtershauser, G. Before enzymes and templates: theory of surface metabolism. *Microbiol Rev* **52**, 452-484, doi:10.1128/mr.52.4.452-484.1988 (1988).

171 Lane, N. & Martin, W. F. The origin of membrane bioenergetics. *Cell* **151**, 1406-1416, doi:10.1016/j.cell.2012.11.050 (2012).

172 Herter, S., Fuchs, G., Bacher, A. & Eisenreich, W. A bicyclic autotrophic $CO2$ fixation pathway in Chloroflexus aurantiacus. *J Biol Chem* **277**, 20277-20283, doi:10.1074/jbc.M201030200 (2002).

173 Smith, E. & Morowitz, H. J. Universality in intermediary metabolism. *Proc Natl Acad Sci U S A* **101**, 13168-13173, doi:10.1073/pnas.0404922101 (2004).

174 Orgel, L. E. The implausibility of metabolic cycles on the prebiotic Earth. *PLoS Biol* **6**, e18, doi:10.1371/journal.pbio.0060018 (2008).

175 Joyce, G. F. & Szostak, J. W. Protocells and RNA Self-Replication. *Cold Spring Harb Perspect Biol* **10**, doi:10.1101/cshperspect.a034801 (2018).

176 Rajamani, S. *et al.* Lipid-assisted synthesis of RNA-like polymers from mononucleotides. *Orig Life Evol Biosph* **38**, 57-74, doi:10.1007/s11084-007-9113-2 (2008).

Index

D

da Vinci, Leonardo 78
Dale, Henry 70-1, 99
Dale, Henry Hallett 70, 121
Damadian, Raymond 112-14
David, Jacques-Louis 12
de Meyer, Jean 172
Deamer, David 209
Dean, Robert 148
death 58, 66, 92, 123, 220-1, 224
degeneration 172-3
dehydration 124, 145
demon 192, 197-8
denaturation process 139
dendrites 91-2, 98, 100
deoxy nucleotides 202
detoxification 165
diabetes 16, 153, 163, 170-3, 175, 220, 223
diabetes mellitus 171, 222
diarrhoea 145, 171
digestion 14, 138-9, 141-2, 149, 170
digestion process 138, 170
dilation 75, 82
discharging 54, 102
dizziness 58
DNA (deoxyribonucleic acid) 7, 20, 150, 179-80, 182, 184-6, 194-5, 200, 202-4, 206, 208
Dobson, Matthew 171
dogs 141, 155, 164, 171, 173, 177, 194-6
dopamine 105, 118
dopamine transporters 118
double helix 7
Doudna, Jennifer 204, 224
drug pumps 169, 191
drugs 81, 119, 141, 153, 168-9, 177, 183, 186-7, 189-90, 220
    anti-acid 140
    anticancer 187

antidepressant 118
antiparasitic 141
    new 140, 187, 189
    over-the-counter 132
Druker, Brian 187
du Bois-Reymond, Emil 23, 94, 107
du Duve, Christian 160
Dumas, Jean Baptiste 13
duodenum 142
dust 67

E

Eberle, Nepomuk 139
Eccles, John Carew 99, 101, 121
Edkins, John Sydney 139
EEG (electroencephalogram) 96, 110, 219
Efstratiadis, Argiris 180-1
Eggleton, Grace 41
Eggleton, Philip 41
electrical signals 90, 93, 109
electricity 4, 90-1, 98, 101, 105, 129, 201
electrodes 97
electron configuration 8
electron microscopy 47, 65
electrons 8, 44, 48-50, 52-4, 59-60, 62-3, 89, 150, 184, 207
    removal of 43, 48, 52
    storing 53
Embden, Gustav 19, 21, 25-7, 30, 32, 36, 41, 216
Embden-Meyerhof-Parnas pathway 26, 164
Ember Therapeutics 131
embryo development 184
emotions 1, 192
endothelial cells 92
energetics 19
energy 1, 3, 7, 48, 129, 182, 197, 199, 202, 206, 213

Fiske, Cyrus Hartwell   27-8, 41
Fletcher, Walter Morley   24
fMRI (functional magnetic resonance
      imaging)   110, 220
foetuses   83
folate   28
food   10, 13, 78, 123-4, 126, 136-
      8, 146, 154, 158-9, 169, 171,
      190, 201
   digested   136, 140
   fatty   132
food digestion   136, 171
foodstuff   13-15, 18, 33, 48, 50
   burning   10
formaldehyde   206
Frankenstein (Shelley)   90
Freda (Lipmann's wife)   34-5
frogs   39, 69, 216
fructose   144
fuels   44-5, 126, 130, 133, 176, 218
   fossil   48

G

Galenus, Aelius   78
Galvani, Luigi   90
gases   5, 11, 206
gastric juice   14, 138-9, 142
GDP (guanosine diphosphate)   188
gene therapy   189
Genentech   180-2, 223
genes   83, 108, 148, 176, 179, 184-5,
      189-90, 200, 202-3, 207
   human   180-1, 223
genome   148, 184, 186, 196
Gilbert, Walter   179-81
Gilman, Alfred   108, 121
glands   134, 136
   hibernating   129, 131
   sweat   136
Gleevec   187, 190
   developers of   187

glomerulus   151-3
glucagon   134, 160-2
gluconeogenesis   163
glucose   24, 41, 44-5, 50, 58, 73, 81, 91,
      106, 111, 119, 126-7, 133, 143-6,
      148, 152-3, 155-6, 158-65, 170-
      1, 175-7, 183-4, 215, 222-3
   breakdown of   30, 164, 183-4
   polymer of   24, 73
   radioactive   147
   removal of   144, 153, 171
   sparing   128
   spillover of   153, 174
   storing   159, 175
glucose consumption   111
glucose generation   127, 165
glucose metabolism
   controlled   172
   elucidation of   24, 26
glucose molecules   24
glucose sensor   159
glutamate   105-6, 120
glutamate levels   120
glycerol   126-7, 134, 164
glycine   207
glycogen   23-5, 30, 41, 45, 73, 75-6, 87,
      99, 111, 119, 124, 126, 128, 132,
      155-6, 163-4, 175, 216
   breakdown of   25-6, 115, 156,
      161, 185
   endogenous   45
   synthesises   158
glycogen account   158-60
glycogen metabolism   155
glycogen reserves   46
glycolysis   26, 28, 30-2, 36, 39-44, 47,
      58, 62, 73
glycolysis pathway   26, 41-2
glycolytic pathways   36
Gmelin, Leopold   138
Goeddel, David   181

nucleotides 2, 7, 155-6, 161, 182, 202,
205-7, 210
    polymerising 209-10
    stitching 204
nutrient absorption 144-6, 148
nutrient fragments 49-52, 134, 150
nutrients 10, 13-14, 43, 46-50, 57, 74,
77-80, 87, 90, 92, 132, 137, 143,
145-7, 150, 152-5, 158, 162, 164,
168, 175, 194, 196, 208
    absorbed 80
    breakdown of 34, 53, 206
    conversion of 13-14
    major 132
nutrition 4, 15, 141

O

obesity 125, 171, 220-1
Ochoa, Severo 20, 36, 159
Ohsumi, Yoshinori 109
oligodendrocytes 91-2
omeprazole 141
organelles 47
organisms 8, 13-15, 47, 137, 192, 200,
203, 209, 217
    ancient 203
    complex 202, 212
    functional 205
    living 14, 17, 207
organs 14, 47, 78-9, 90, 124, 128, 149,
154-5, 164, 183-4
ornithine 166, 168
Oro, Joan 206
osmolarity 152
osmosis 198
osmotic gradients 151
ouabain 104
oxidation 11, 13, 48, 106, 210, 217-18
oxygen:
    absence of 23-4, 35, 60, 88, 207
    attaching 48, 165

    binding of 87
    sufficient 73
    transport of 13, 25, 48, 219
oxygen atoms 2-3, 47-8
oxygen binding 26, 82-3, 86-7
oxygen consumption 111
oxygen exchange 77
oxygen levels 84
oxygen pressure 82-3
oxygen supply 87

P

P-gp (P-glycoprotein) 190
pacemaker cells 81
paddle wheels 53-5, 59-60, 62, 129
pain 45, 74, 82, 88, 102, 106
pancreas 80, 142, 154, 171-4, 176-7,
180-1
pancreas duct 173
pancreas extracts 172-3
pancreatic juices 142-3
pancreatic stone, rare 173
Parnas, Jakub Karol 21, 26, 32, 41
Pasteur, Louis 16
Patapoutian, Ardem 102, 122
Pauling, Linus 85, 110
pearls 137-9, 141-3, 166, 179-80,
200, 204
    strings of 141, 143, 182
peloton analogy 126, 133
pelotons 49, 126-7, 134
pepsin 14, 139, 141-3, 178, 221
pepsin generator 139
pepsinogen 139
perception 53, 109
periodicity 65
Perutz, Max 84-6, 219
petrol 4, 10-11, 134
Pfizer 28
Pflüger, Eduard 13
pH gradients 208

respiration 59-61, 63, 73, 75

respiration enzyme 59

'Revolution in Muscle Physiology' (Hill) 27

ribosome 194, 200, 203-5, 224

ribozymes 204-5, 224

Richards, Alfred Newton 152

Riggs, Arthur 181

Riordan, Jack 190

RNA (ribonucleic acid) 7, 20, 150, 203-8, 210, 224

RNA molecule 204-5, 208

Rodbell, Martin 108, 121

Rorsman, P. 176, 223

Rosbash, Michael 102, 122

Rutter, William 179-80

S

Sachs, George 140

Sakmann, Bert 97, 121

salt 28, 145, 151-2, 171

salt concentration 152

salt molecules 151

salt solution 69-70

Sanger, Frederick 178-9

Sawyer, Charles 187

scar tissue 184

Schnitger, Heinrich 58

Schrödinger, Erwin 192-3

Schwann, Theodor 14-15, 17, 139

science 10, 19-21, 32, 126, 212, 215-16, 220, 222-4

scientific community, in Berlin 36

scientists 9, 20, 29, 51, 138, 159, 192-3

Second World War 19, 21, 69

secretin 173

secretion 107, 176

Sharpey-Schafer, Edward Albert 172

sheep 94-7, 100, 102, 104, 145-7, 196-9

sheepdogs 145-7

Shelley, Mary 90

Sherrington, Charles Scott 96, 98, 110, 121

signal transduction 121

silver stain 92-3

Skou, Jens Christian 103-4

sleep 102, 118-19, 132, 220

sleep deprivation 118

small intestine 136-7, 142

Smith, Robert 131

soap bubbles 54, 62, 108-9, 193

sodium 104, 120, 142, 145, 148, 153

sodium fluoride 25

sodium ions 102-4, 142, 144-6, 148, 151-2, 193

sodium pump 80, 102-5, 140, 142, 144-9, 151, 193

somatostatin 181

Spiegelman, Bruce 130

splanchnic bed 80, 154

spleen 80, 154

starvation 123, 125-7, 221

steam engines 5, 197, 201

stents 82

Stickland, L. H. 207

stomach 80, 133, 136, 138-43, 154

stomach acid 142

Straub, Bruno Ferenc 71

Subbarow, Yellapragada 27-8, 41

sugar compounds 31

sugar cubes 7, 22-3

sugar metabolism 21, 25, 28

sugar molecules 25

sugar-phosphate esters 41

sugar phosphates 25, 29

sugars 2, 14-16, 19, 23, 25, 28, 30, 39, 50, 74, 81, 137-9, 143-4, 147-8, 154-5, 171, 194, 206, 208, 221

breakdown of 18, 21, 23, 26, 31-2, 39, 41

incubated radioactive 147

www.ingramcontent.com/pod-product-compliance
Lightning Source LLC
Chambersburg PA
CBHW021356210526
45463CB00001B/113